Preface

This book is intended for use at the laboratory bench, to guide the student in the interpretation of his own specimens: it is not a textbook and thus does not contain theoretical information. As an introduction to the study of chordate structure, this book should be used in conjunction with those on histology and embryology already published in this series.

All the photographs and drawings have been specially made for the book, and a wide variety of specimens has been included. Nowadays the student has to deal with many kinds of preparation in addition to the museum mount and permanent microscope slide, and we have used some commercial mounts as well as our own dissections. In some cases, especially dogfish as a basic chordate type, we have used dissections, commercial mounts, infrared photography of living embryos, flash photography of living embryos, permanent microscope preparations, freshly-cut steaks of preserved fish—all to elucidate that structure without which a knowledge of function cannot be acquired. This wide variety of material should also be investigated by the student, for no one kind of specimen can give adequate information. We have been fortunate in being able to augment our own range of specimens by borrowing from colleagues—Savile Bradbury, Frank Cox, Bill Freeman and Anne Terry have all been most helpful in this regard, and we owe them sincere thanks for their willingness to co-operate.

We have been equally fortunate in commercial sources. To Gerrard and Company we owe considerable thanks for providing on loan, often at considerable inconvenience to themselves, some of their superb specimens: numbers 48, 49, 50, 51, 67, 70, 71, 75, 76, 103, 104, and 106. To Philip Harris Biological we are equally indebted: specimens 27, 54, 55, 69, 72, 73, 87, 88, 90, 91, 93, and 96 came directly from them, as also did most of the animals we dissected for ourselves. There is no doubt that the student uses commercial preparations in wide variety nowadays, and we are most grateful for the chance to include these excellent examples from our friends in the biological supply houses.

To Bill Freeman we owe an especial debt of gratitude.

He has read the work in manuscript with his usual total thoroughness, and to him we give thanks not only for saving us from error, but also for encouraging us in the work.

B. B.
P. H. M.

January 1977

By the same authors
An Atlas of Plant Structure: Volume 1
An Atlas of Plant Structure: Volume 2

By W. H. Freeman and Brian Bracegirdle
An Atlas of Invertebrate Structure
An Atlas of Embryology
An Atlas of Histology
An Advanced Atlas of Histology

Colour Transparencies for Projection

Every photograph in this book is available as a 2×2 colour slide for projection from Philip Harris Biological Ltd., Oldmixon, Weston-super-Mare, Avon.

Each original master transparency was made at the same time as the negative for the corresponding picture in this book, exclusively for this Company. The authors recommend these slides for their quality and moderate cost for the teaching of chordate structure, especially in conjunction with this book.

AN ATLAS OF
Chordate
Structure

Brian Bracegirdle BSc PhD FIBiol FRPS

and

Patricia H Miles MSc MIBiol ARPS

Heinemann Educational Books
London

Heinemann Educational Books Ltd
22 Bedford Square, London WC1B 3HH
LONDON EDINBURGH MELBOURNE AUCKLAND
HONG KONG SINGAPORE KUALA LUMPUR NEW DELHI
IBADAN NAIROBI JOHANNESBURG
EXETER (NH) KINGSTON PORT OF SPAIN

ISBN 0 435 60316 7

Printed and bound in Great Britain at
The University Press, Oxford

CONTENTS

OSTEICHTHYES

AMPHIBIA

REPTILIA

AVES

MAMMALIA

DERMATOLOGY

OSTEOLOGY

1. *Clavelina*, E. (Mag. ×20)

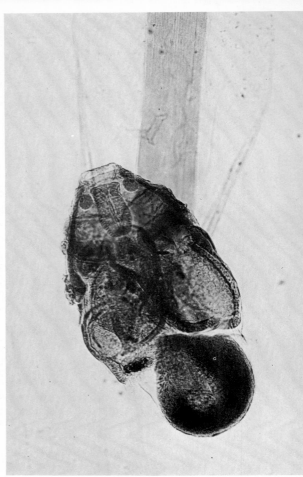

2. *Oikopleura* larva, anterior E. (Mag. ×75)

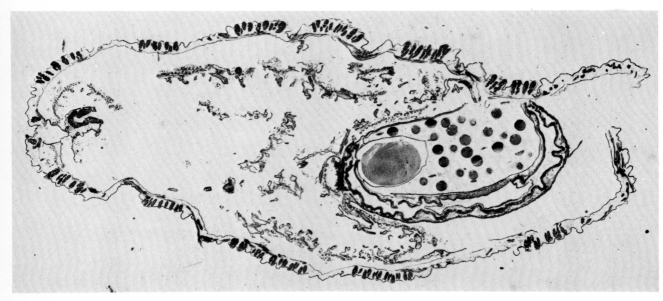

3. *Ciona*, TS oblique. (Mag. ×45)

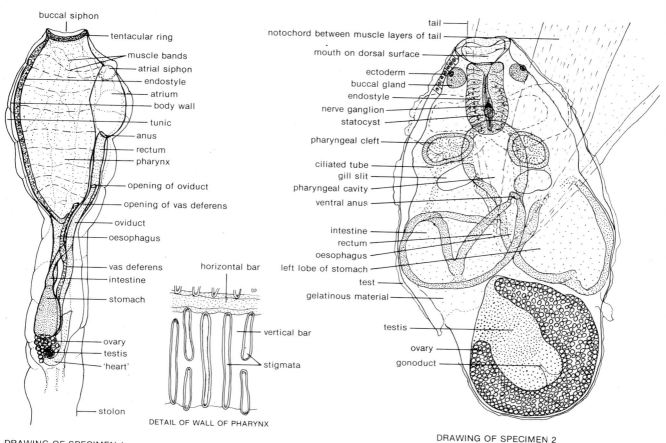

buccal siphon
tentacular ring
muscle bands
atrial siphon
endostyle
atrium
body wall
tunic
anus
rectum
pharynx
opening of oviduct
opening of vas deferens
oviduct
oesophagus
vas deferens
intestine
stomach
ovary
testis
'heart'
stolon

DRAWING OF SPECIMEN 1

horizontal bar
vertical bar
stigmata

DETAIL OF WALL OF PHARYNX

tail
notochord between muscle layers of tail
mouth on dorsal surface
ectoderm
buccal gland
endostyle
nerve ganglion
statocyst
pharyngeal cleft
ciliated tube
gill slit
pharyngeal cavity
ventral anus
intestine
rectum
oesophagus
left lobe of stomach
test
gelatinous material
testis
ovary
gonoduct

DRAWING OF SPECIMEN 2

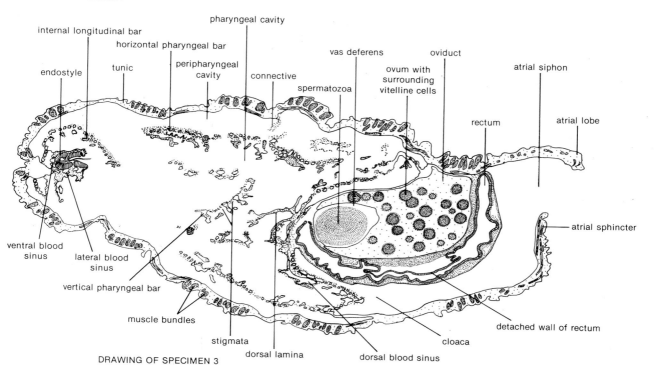

internal longitudinal bar
horizontal pharyngeal bar
pharyngeal cavity
peripharyngeal cavity
connective
vas deferens
ovum with surrounding vitelline cells
oviduct
atrial siphon
endostyle
tunic
spermatozoa
rectum
atrial lobe
atrial sphincter
ventral blood sinus
lateral blood sinus
vertical pharyngeal bar
muscle bundles
stigmata
dorsal lamina
dorsal blood sinus
cloaca
detached wall of rectum

DRAWING OF SPECIMEN 3

1

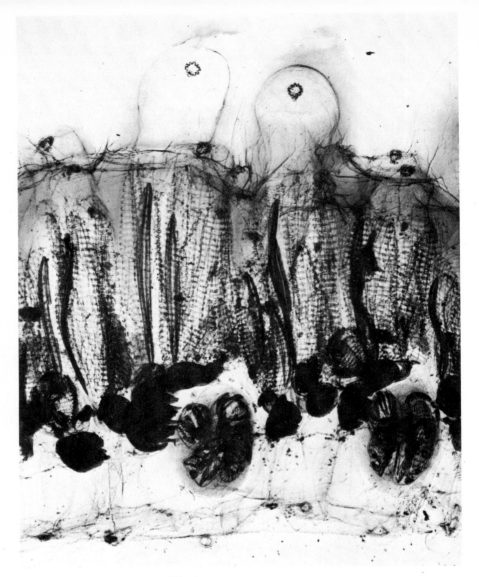

4. *Pyrosoma* colony, E. (Mag. ×30)

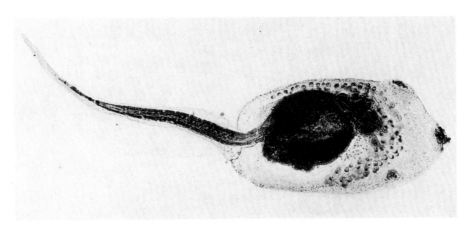

5. Ascidian larva E, anterior. (Mag. ×45)

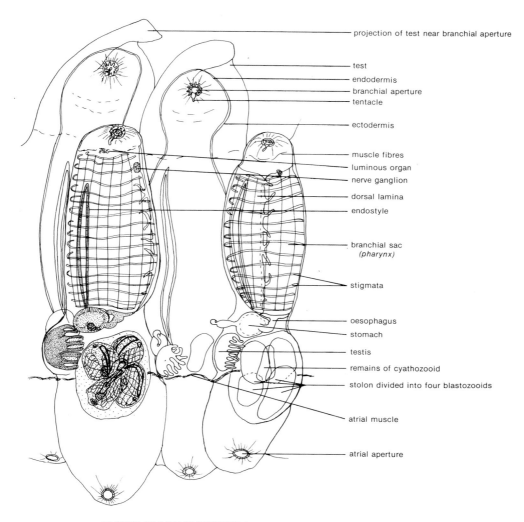

- projection of test near branchial aperture
- test
- endodermis
- branchial aperture
- tentacle
- ectodermis
- muscle fibres
- luminous organ
- nerve ganglion
- dorsal lamina
- endostyle
- branchial sac
  (pharynx)
- stigmata
- oesophagus
- stomach
- testis
- remains of cyathozooid
- stolon divided into four blastozooids
- atrial muscle
- atrial aperture

DRAWING OF PART OF SPECIMEN 4

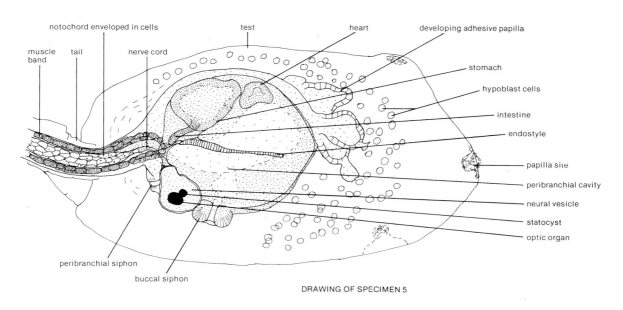

- notochord enveloped in cells
- test
- heart
- developing adhesive papilla
- muscle band
- tail
- nerve cord
- stomach
- hypoblast cells
- intestine
- endostyle
- papilla site
- peribranchial cavity
- neural vesicle
- statocyst
- optic organ
- peribranchial siphon
- buccal siphon

DRAWING OF SPECIMEN 5

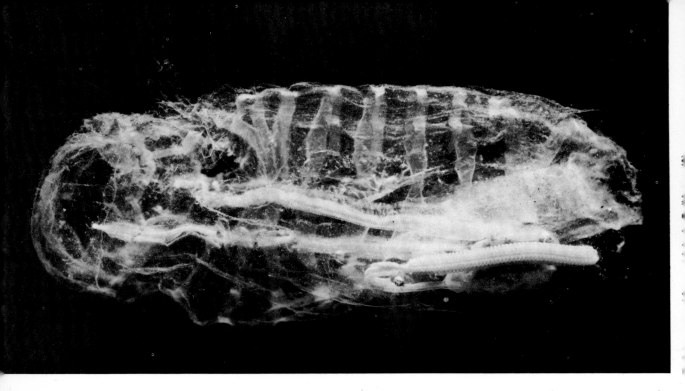

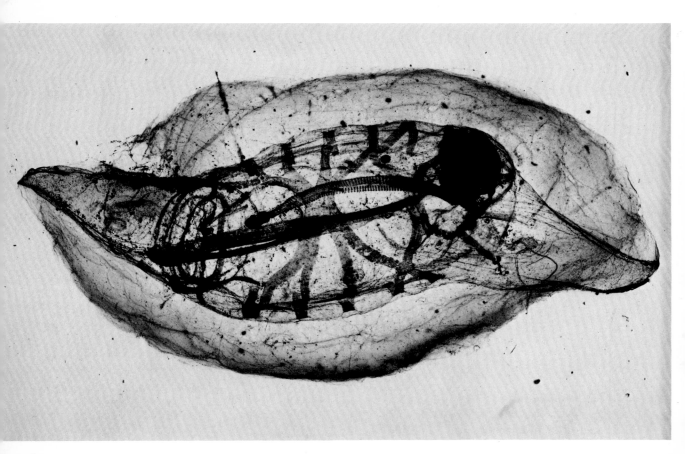

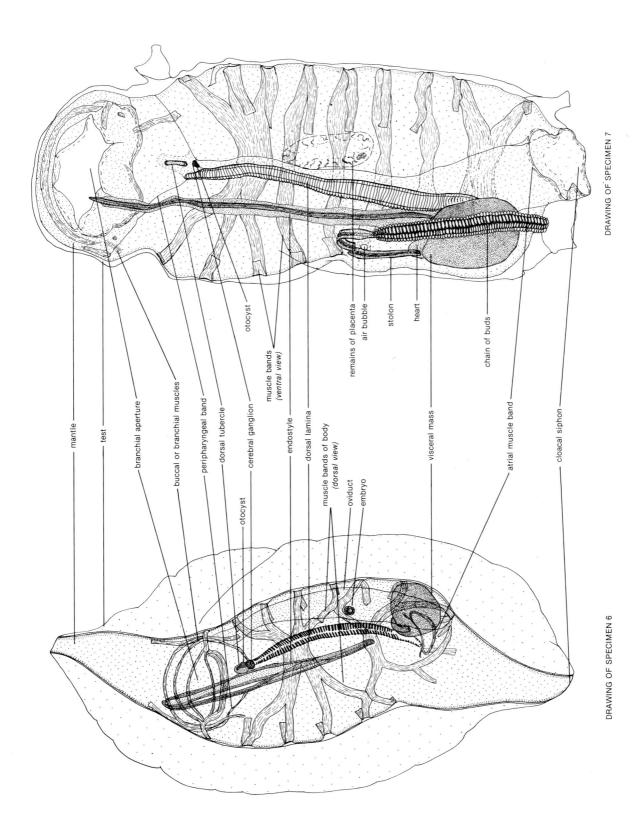

DRAWING OF SPECIMEN 7

DRAWING OF SPECIMEN 6

mantle

test

branchial aperture

buccal or branchial muscles

peripharyngeal band

dorsal tubercle

cerebral ganglion

otocyst

otocyst

endostyle

dorsal lamina

muscle bands
(ventral view)

muscle bands of body
(dorsal view)

oviduct

embryo

remains of placenta

air bubble

stolon

heart

visceral mass

chain of buds

atrial muscle band

cloacal siphon

5

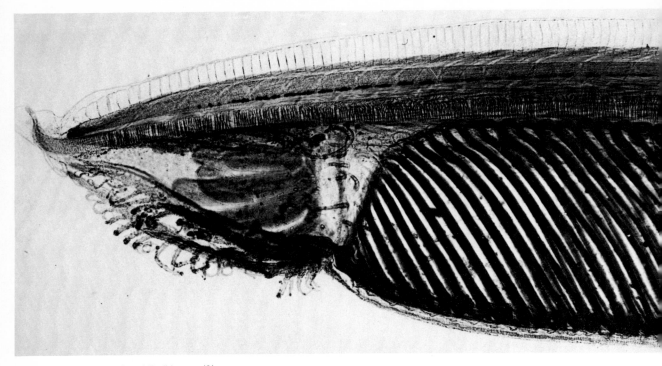

8. *Branchiostoma*, oral hood E. (Mag. ×42)

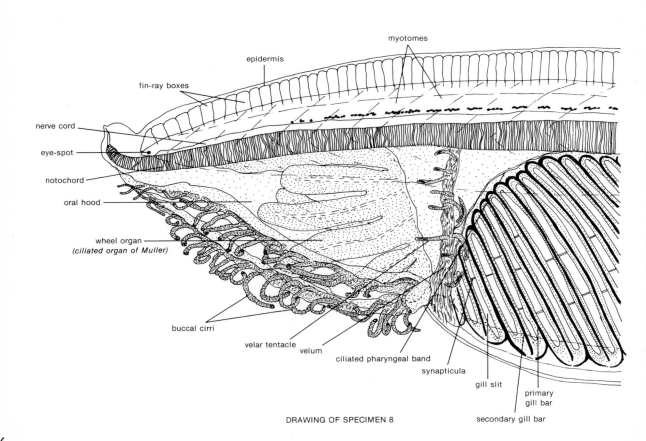

DRAWING OF SPECIMEN 8

6

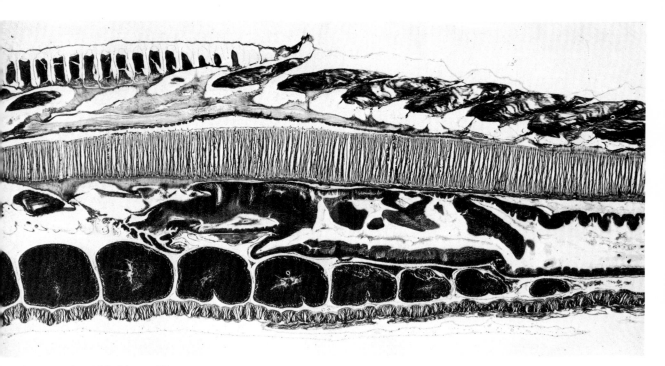

Branchiostoma, detail LS. (Mag. ×35)

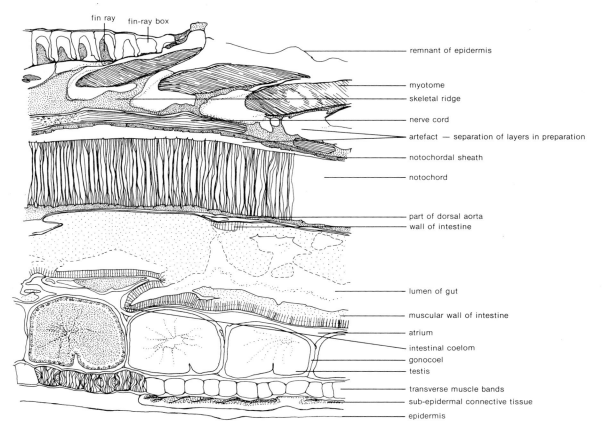

fin ray fin-ray box

remnant of epidermis

myotome
skeletal ridge
nerve cord
artefact — separation of layers in preparation

notochordal sheath
notochord

part of dorsal aorta
wall of intestine

lumen of gut

muscular wall of intestine
atrium
intestinal coelom
gonocoel
testis
transverse muscle bands
sub-epidermal connective tissue
epidermis

DRAWING OF SPECIMEN 9

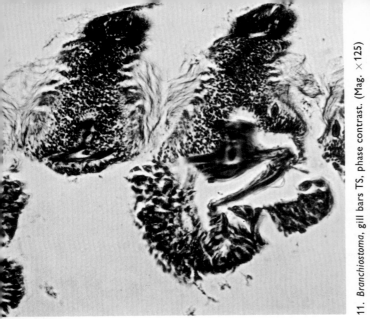
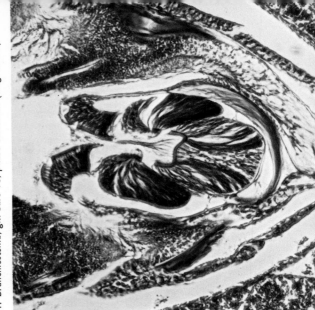

11. *Branchiostoma*, gill bars TS, phase contrast. (Mag. ×125)

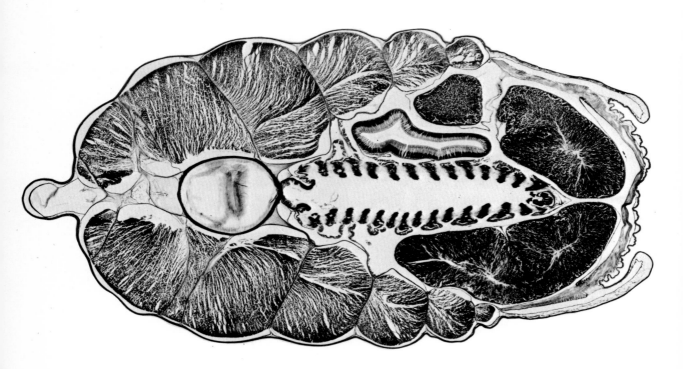

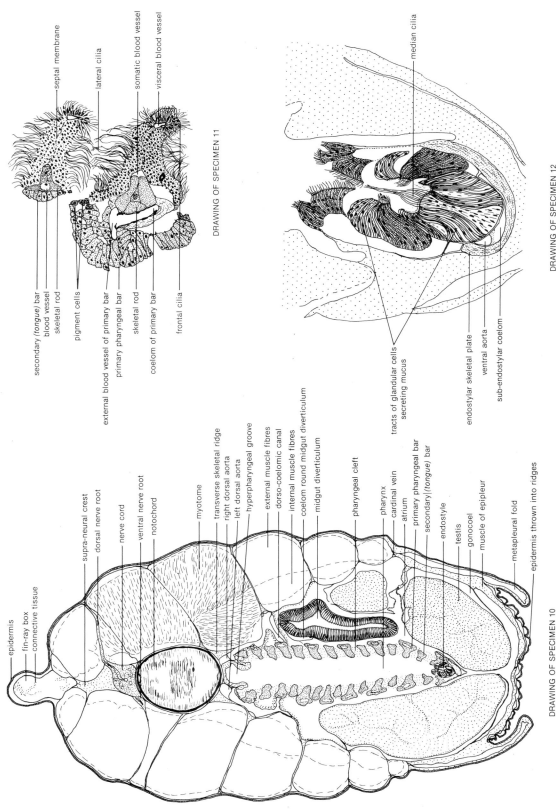

septal membrane
lateral cilia
somatic blood vessel
visceral blood vessel

DRAWING OF SPECIMEN 11

secondary (tongue) bar
blood vessel
skeletal rod
pigment cells
external blood vessel of primary bar
primary pharyngeal bar
skeletal rod
coelom of primary bar
frontal cilia

median cilia

DRAWING OF SPECIMEN 12

tracts of glandular cells
secreting mucus
endostylar skeletal plate
ventral aorta
sub-endostylar coelom

epidermis
fin-ray box
connective tissue

supra-neural crest
dorsal nerve root
nerve cord
ventral nerve root
notochord

myotome
transverse skeletal ridge
right dorsal aorta
left dorsal aorta
hyperpharyngeal groove
external muscle fibres
dorso-coelomic canal
internal muscle fibres
coelom round midgut diverticulum
midgut diverticulum

pharyngeal cleft

pharynx
cardinal vein
atrium
primary pharyngeal bar
secondary (tongue) bar
endostyle
testis
gonocoel
muscle of epipleur
metapleural fold
epidermis thrown into ridges

DRAWING OF SPECIMEN 10

9

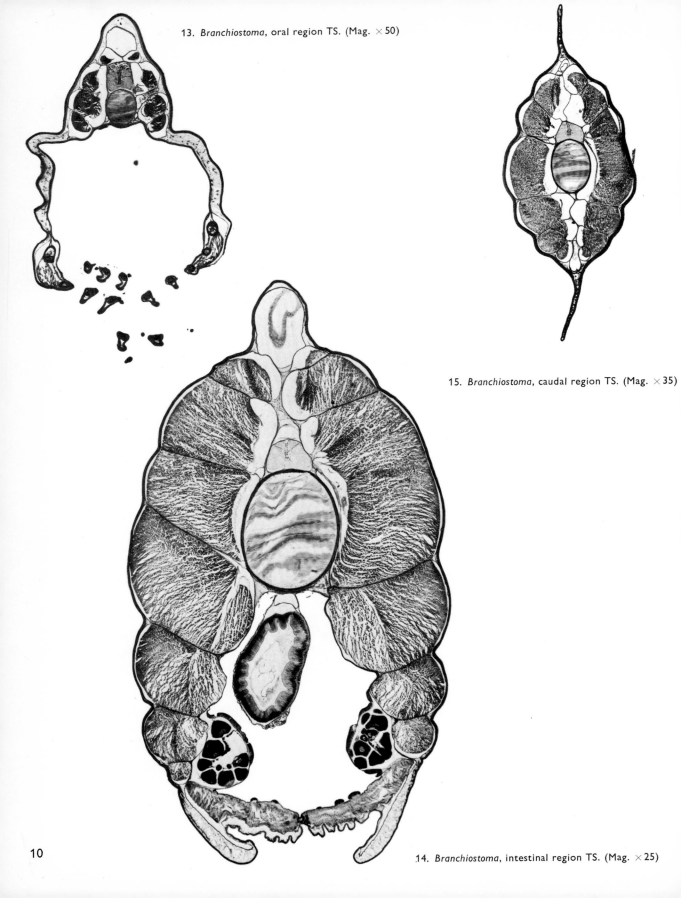

13. *Branchiostoma*, oral region TS. (Mag. ×50)

15. *Branchiostoma*, caudal region TS. (Mag. ×35)

14. *Branchiostoma*, intestinal region TS. (Mag. ×25)

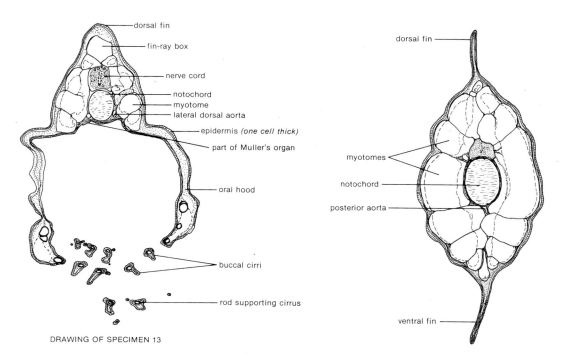

dorsal fin

fin-ray box

nerve cord

notochord
myotome
lateral dorsal aorta

epidermis *(one cell thick)*

part of Muller's organ

oral hood

buccal cirri

rod supporting cirrus

DRAWING OF SPECIMEN 13

dorsal fin

myotomes

notochord

posterior aorta

ventral fin

DRAWING OF SPECIMEN 15

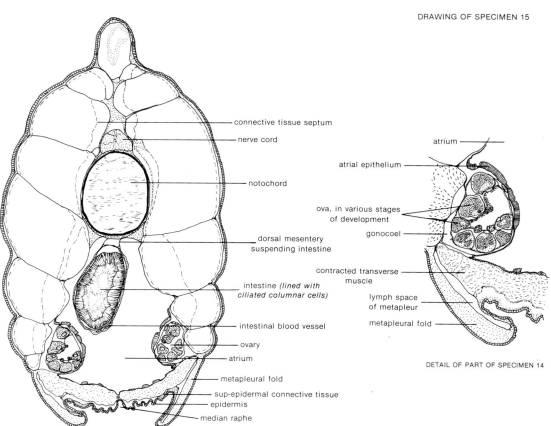

connective tissue septum

nerve cord

notochord

dorsal mesentery
suspending intestine

intestine *(lined with
ciliated columnar cells)*

intestinal blood vessel

ovary

atrium

metapleural fold

sup-epidermal connective tissue

epidermis

median raphe

DRAWING OF SPECIMEN 14

atrium

atrial epithelium

ova, in various stages
of development

gonocoel

contracted transverse
muscle

lymph space
of metapleur

metapleural fold

DETAIL OF PART OF SPECIMEN 14

16. *Petromyzon*, dissection of anterior. (Mag. ×2)

17. *Petromyzon*, head, sagittal LS. (Mag. ×10)

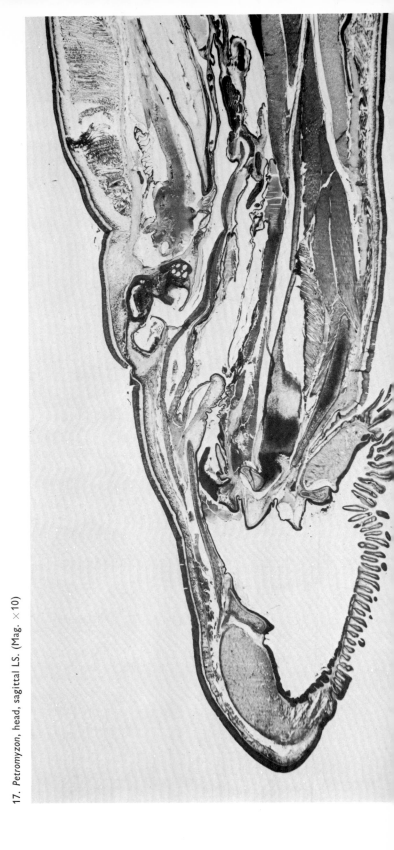

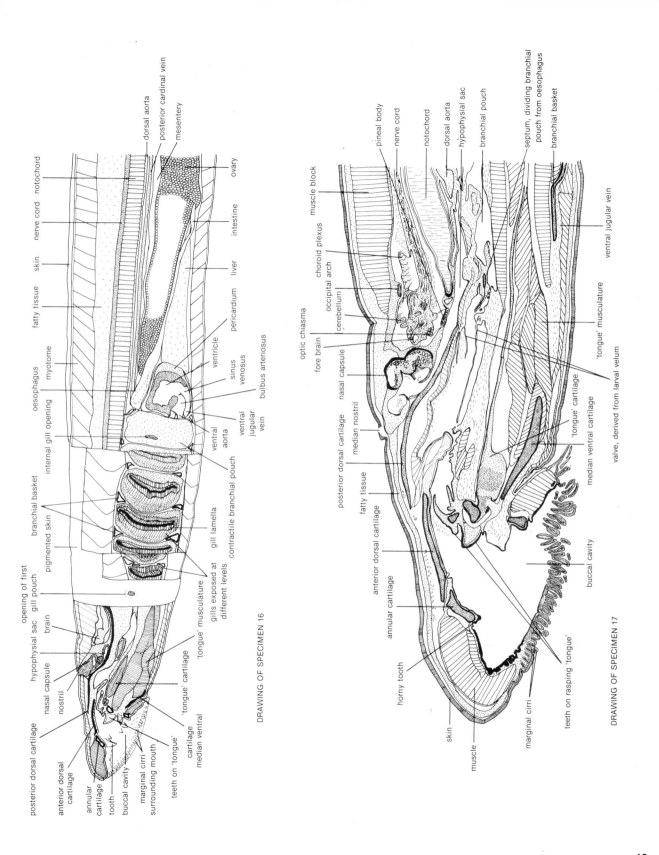

posterior dorsal cartilage

anterior dorsal cartilage

nostril

nasal capsule

brain

hypophysial sac

opening of first gill pouch

pigmented skin

branchial basket

internal gill opening

oesophagus

myotome

fatty tissue

skin

nerve cord

notochord

dorsal aorta

posterior cardinal vein

mesentery

ovary

intestine

liver

pericardium

ventricle

sinus venosus

bulbus arteriosus

ventral aorta

jugular vein

contractile branchial pouch

gill lamella

gills exposed at different levels

'tongue' musculature

'tongue' cartilage

median ventral cartilage

teeth on 'tongue'

marginal cirri surrounding mouth

buccal cavity

tooth

annular cartilage

DRAWING OF SPECIMEN 16

optic chiasma

fore brain

choroid plexus

muscle block

pineal body

nerve cord

notochord

dorsal aorta

hypophysial sac

branchial pouch

septum, dividing branchial pouch from oesophagus

branchial basket

ventral jugular vein

'tongue' musculature

valve, derived from larval velum

median ventral cartilage

'tongue' cartilage

buccal cavity

teeth on rasping 'tongue'

marginal cirri

muscle

skin

horny tooth

annular cartilage

anterior dorsal cartilage

fatty tissue

posterior dorsal cartilage

median nostril

nasal capsule

occipital arch

cerebellum

DRAWING OF SPECIMEN 17

13

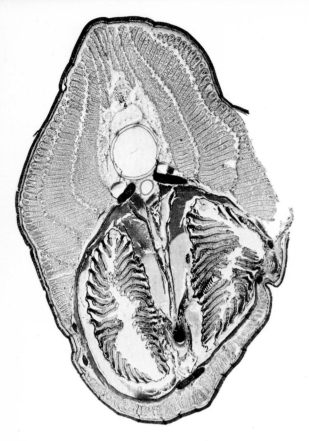

18. *Petromyzon*, gill region TS. (Mag. ×6)

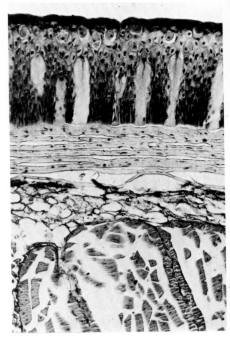

19. *Petromyzon*, skin VS. (Mag. ×245)

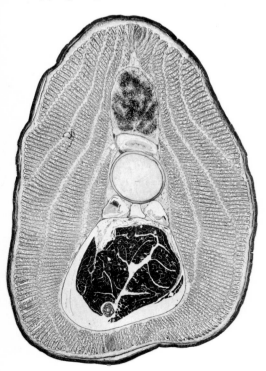

20. *Petromyzon*, liver region TS. (Mag. ×6)

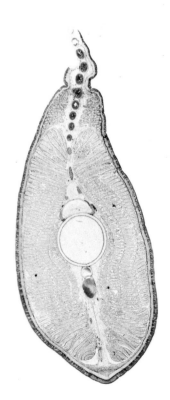

21. *Petromyzon*, caudal region TS. (Mag. ×6)

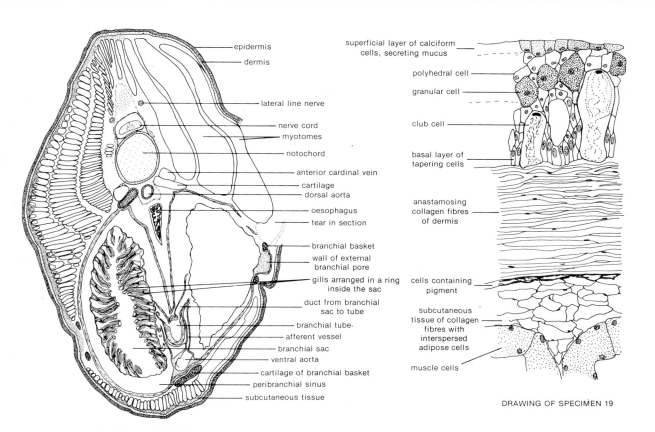

epidermis
dermis

lateral line nerve
nerve cord
myotomes
notochord
anterior cardinal vein
cartilage
dorsal aorta
oesophagus
tear in section
branchial basket
wall of external branchial pore
gills arranged in a ring inside the sac
duct from branchial sac to tube
branchial tube
afferent vessel
branchial sac
ventral aorta
cartilage of branchial basket
peribranchial sinus
subcutaneous tissue

DRAWING OF SPECIMEN 18

superficial layer of calciform cells, secreting mucus
polyhedral cell
granular cell
club cell
basal layer of tapering cells
anastamosing collagen fibres of dermis
cells containing pigment
subcutaneous tissue of collagen fibres with interspersed adipose cells
muscle cells

DRAWING OF SPECIMEN 19

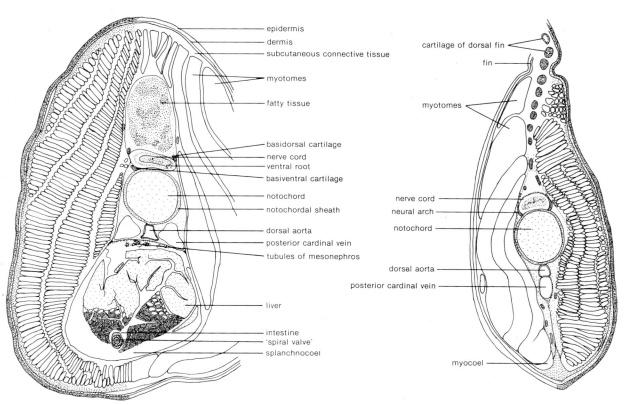

epidermis
dermis
subcutaneous connective tissue
myotomes
fatty tissue
basidorsal cartilage
nerve cord
ventral root
basiventral cartilage
notochord
notochordal sheath
dorsal aorta
posterior cardinal vein
tubules of mesonephros
liver
intestine
'spiral valve'
splanchnocoel

DRAWING OF SPECIMEN 20

cartilage of dorsal fin
fin
myotomes
nerve cord
neural arch
notochord
dorsal aorta
posterior cardinal vein
myocoel

DRAWING OF SPECIMEN 21

15

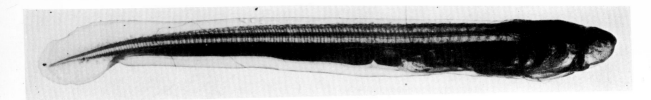

22. *Petromyzon*, ammocoete larva, E. (Mag. ×20)

24. *Petromyzon*, ammocoete larva, liver region TS. (Mag. ×20)

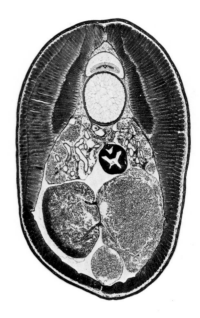

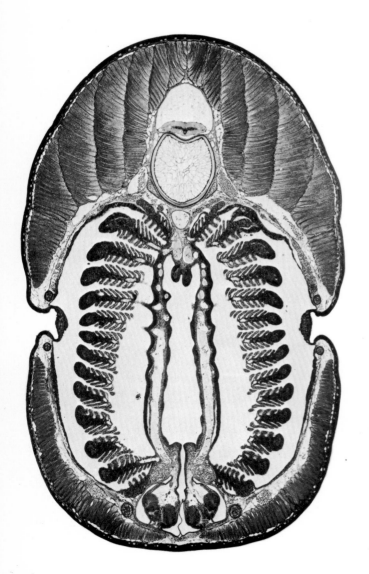

23. *Petromyzon*, ammocoete larva, branchial region TS. (Mag. ×20)

25. *Petromyzon*, ammocoete larva, caudal region TS. (Mag. ×

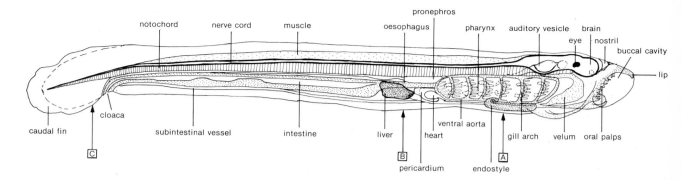

DRAWING OF SPECIMEN 22

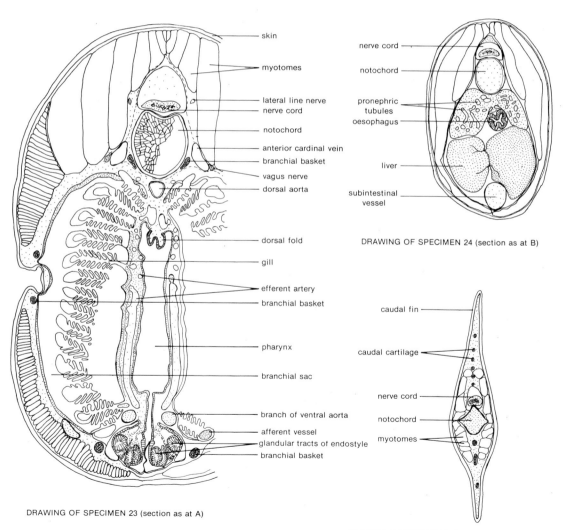

DRAWING OF SPECIMEN 23 (section as at A)

DRAWING OF SPECIMEN 24 (section as at B)

DRAWING OF SPECIMEN 25 (section as at C)

17

18

26. *Scyliorhinus*, general dissection. (Mag. ×0·9)

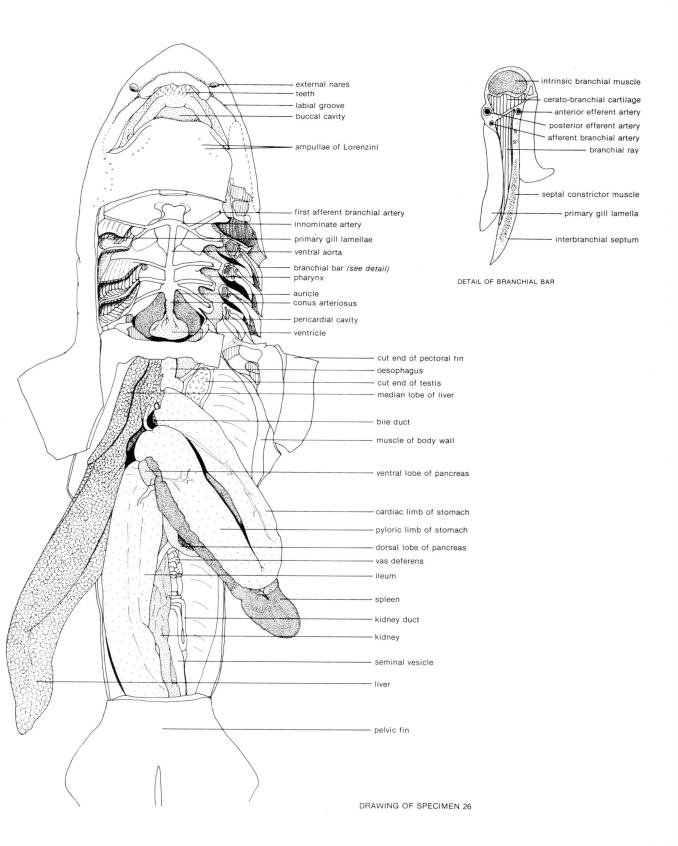

external nares
teeth
labial groove
buccal cavity

ampullae of Lorenzini

intrinsic branchial muscle
cerato-branchial cartilage
anterior efferent artery
posterior efferent artery
afferent branchial artery
branchial ray

septal constrictor muscle

primary gill lamella

interbranchial septum

DETAIL OF BRANCHIAL BAR

first afferent branchial artery
innominate artery
primary gill lamellae
ventral aorta
branchial bar *(see detail)*
pharynx
auricle
conus arteriosus
pericardial cavity
ventricle

cut end of pectoral fin
oesophagus
cut end of testis
median lobe of liver

bile duct

muscle of body wall

ventral lobe of pancreas

cardiac limb of stomach
pyloric limb of stomach
dorsal lobe of pancreas
vas deferens
ileum

spleen

kidney duct

kidney

seminal vesicle

liver

pelvic fin

DRAWING OF SPECIMEN 26

27. *Scyliorhinus*, dissection of heart and blood vessels. (Mag. ×2)

DRAWING OF SPECIMEN 27

Labels (clockwise from top):

external carotid artery
innominate artery
ventral aorta
branchial pouch
afferent branchial arteries II and III
gill bar
conus arteriosus
internal branchial cleft
coronary artery
efferent branchial arteries
ventricle
atrium
sinus venosus
no gill lamellae on posterior face of fifth branchial pouch
position of sinu-atrial aperture
pericardial cavity
pectoral girdle
lateral artery
dorsal aorta
subclavian artery
holobranch
1st epibranchial (efferent) artery
anterior prolongation of dorsal aorta
primary lamellae of anterior hemibranch
mandibular arch
hyoidean artery
internal carotid artery
common carotid artery
spiracle
roof of buccal cavity
connecting branch between trematic arteries
pretrematic artery
post trematic artery
arterial collector loop

1 2 3 4 5

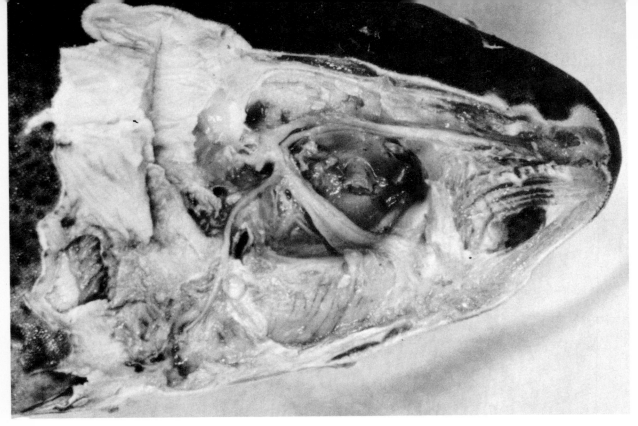

28. *Scyliorhinus*, dissection of cranial nerves 5 and 7. (Mag. ×2)

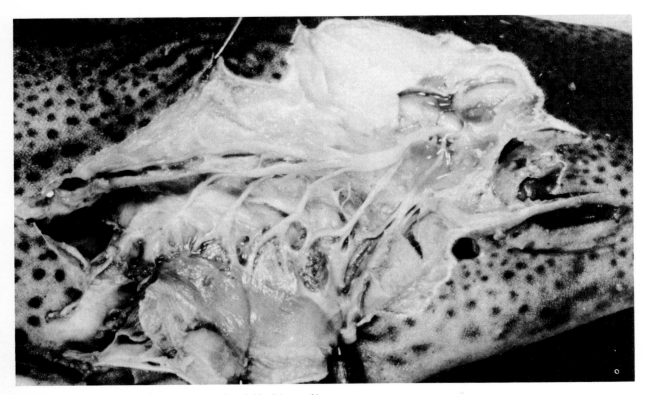

29. *Scyliorhinus*, dissection of cranial nerves 9 and 10. (Mag. ×2)

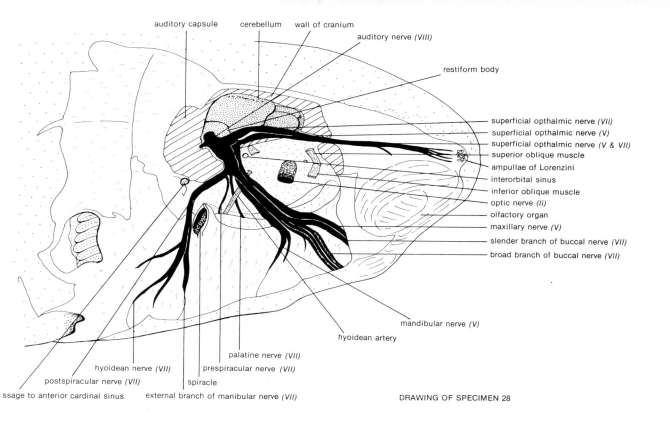

auditory capsule   cerebellum   wall of cranium

auditory nerve (VIII)

restiform body

superficial opthalmic nerve (VII)
superficial opthalmic nerve (V)
superficial opthalmic nerve (V & VII)
superior oblique muscle
ampullae of Lorenzini
interorbital sinus
inferior oblique muscle
optic nerve (II)
olfactory organ
maxillary nerve (V)
slender branch of buccal nerve (VII)
broad branch of buccal nerve (VII)

mandibular nerve (V)

hyoidean artery

palatine nerve (VII)

prespiracular nerve (VII)

hyoidean nerve (VII)

postspiracular nerve (VII)

spiracle

ssage to anterior cardinal sinus

external branch of manibular nerve (VII)

DRAWING OF SPECIMEN 28

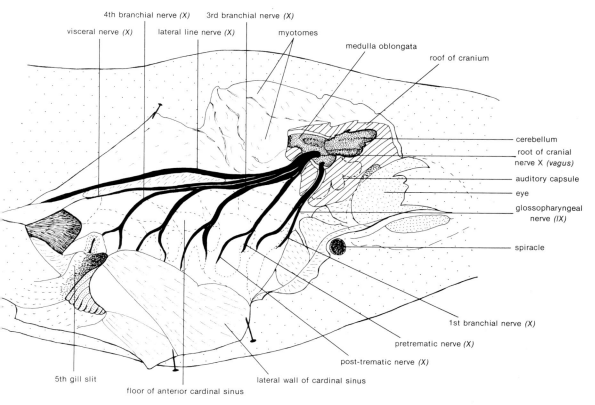

4th branchial nerve (X)   3rd branchial nerve (X)

visceral nerve (X)   lateral line nerve (X)   myotomes

medulla oblongata

roof of cranium

cerebellum
root of cranial
nerve X (vagus)
auditory capsule
eye
glossopharyngeal
nerve (IX)

spiracle

1st branchial nerve (X)

pretrematic nerve (X)

post-trematic nerve (X)

5th gill slit

lateral wall of cardinal sinus

floor of anterior cardinal sinus

DRAWING OF SPECIMEN 29

23

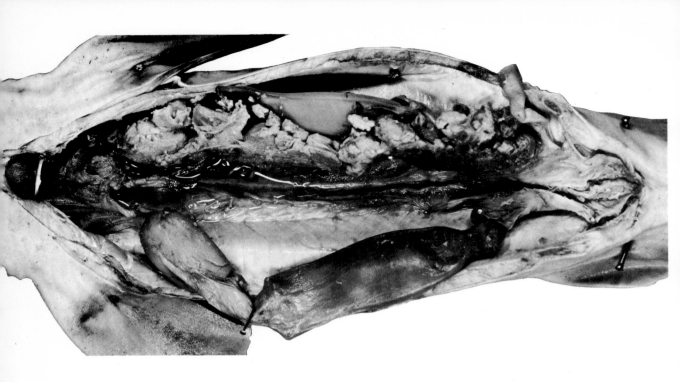

24

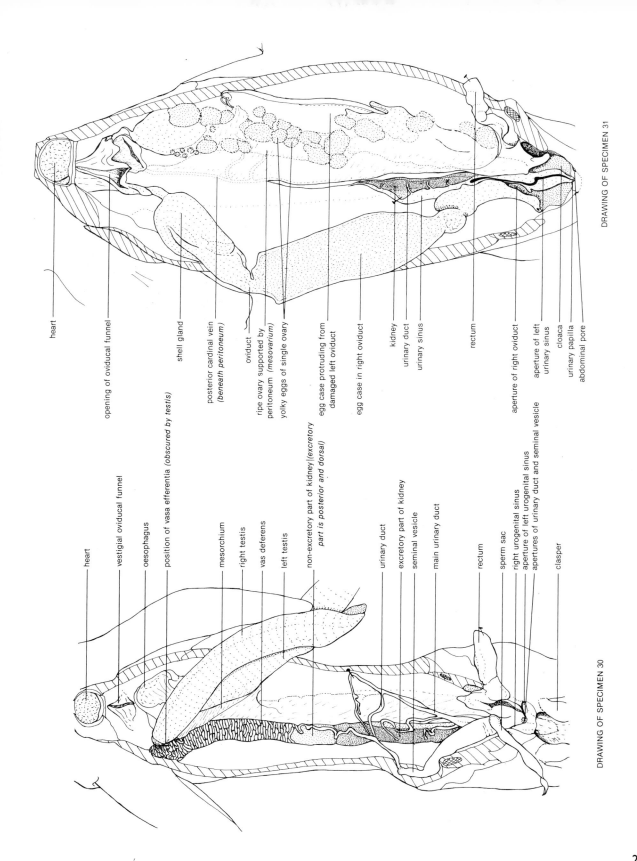

heart

opening of oviducal funnel

shell gland

posterior cardinal vein (beneath peritoneum)

oviduct

ripe ovary supported by peritoneum (mesovarium)

yolky eggs of single ovary

egg case protruding from damaged left oviduct

egg case in right oviduct

kidney

urinary duct

urinary sinus

rectum

aperture of right oviduct

aperture of left urinary sinus

cloaca

urinary papilla

abdominal pore

DRAWING OF SPECIMEN 31

heart

vestigial oviducal funnel

oesophagus

position of vasa efferentia (obscured by testis)

mesorchium

right testis

vas deferens

left testis

non-excretory part of kidney (excretory part is posterior and dorsal)

urinary duct

excretory part of kidney

seminal vesicle

main urinary duct

rectum

sperm sac

right urogenital sinus

aperture of left urogenital sinus

apertures of urinary duct and seminal vesicle

clasper

DRAWING OF SPECIMEN 30

25

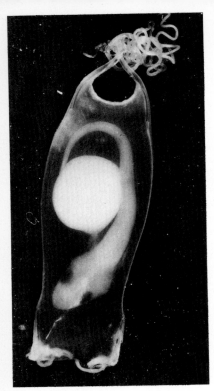

32. *Scyliorhinus*, living embryo in egg case ('mermaid's purse'), photographed by infra-red. (Mag. ×1)

33. *Scyliorhinus*, living embryo and yolk sac. (Mag. ×3)

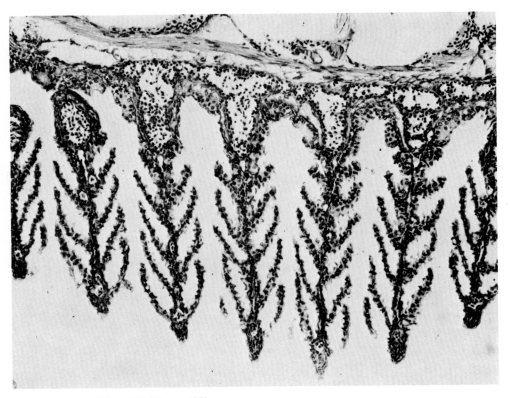

34. *Scyliorhinus*, gill bars TS. (Mag. ×125)

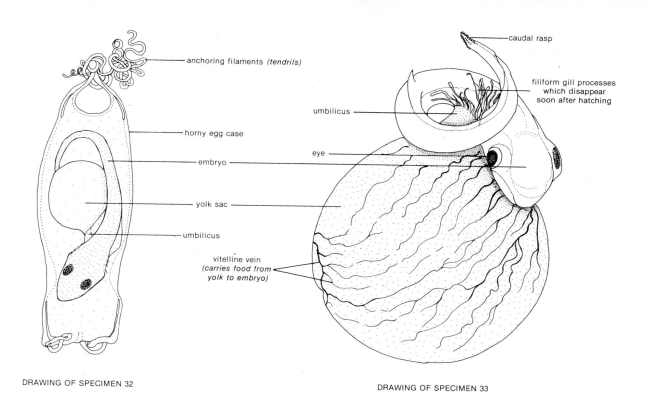

anchoring filaments *(tendrils)*

horny egg case

embryo

yolk sac

umbilicus

DRAWING OF SPECIMEN 32

caudal rasp

filiform gill processes
which disappear
soon after hatching

umbilicus

eye

vitelline vein
*(carries food from
yolk to embryo)*

DRAWING OF SPECIMEN 33

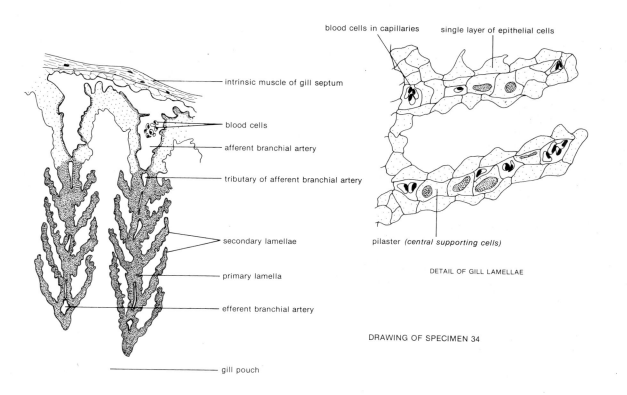

intrinsic muscle of gill septum

blood cells

afferent branchial artery

tributary of afferent branchial artery

secondary lamellae

primary lamella

efferent branchial artery

gill pouch

blood cells in capillaries   single layer of epithelial cells

pilaster *(central supporting cells)*

DETAIL OF GILL LAMELLAE

DRAWING OF SPECIMEN 34

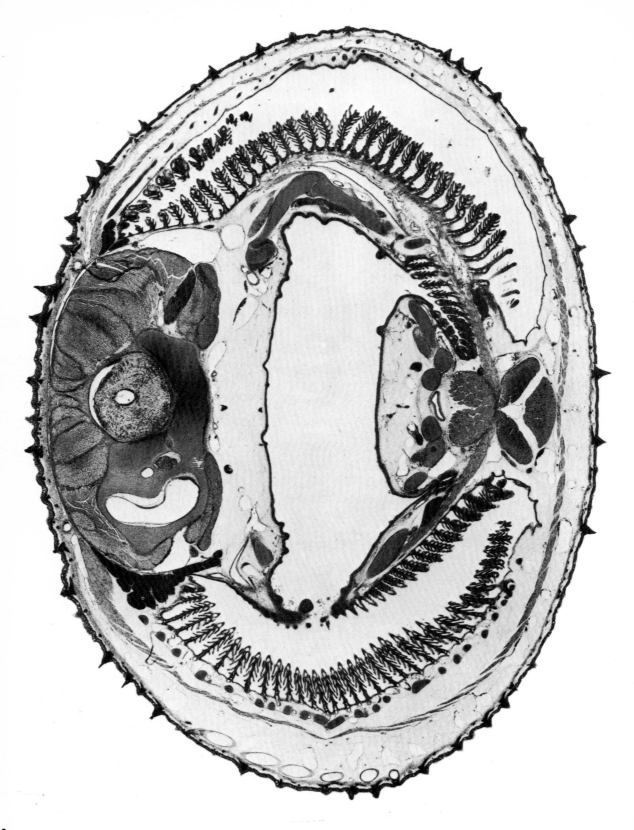

28

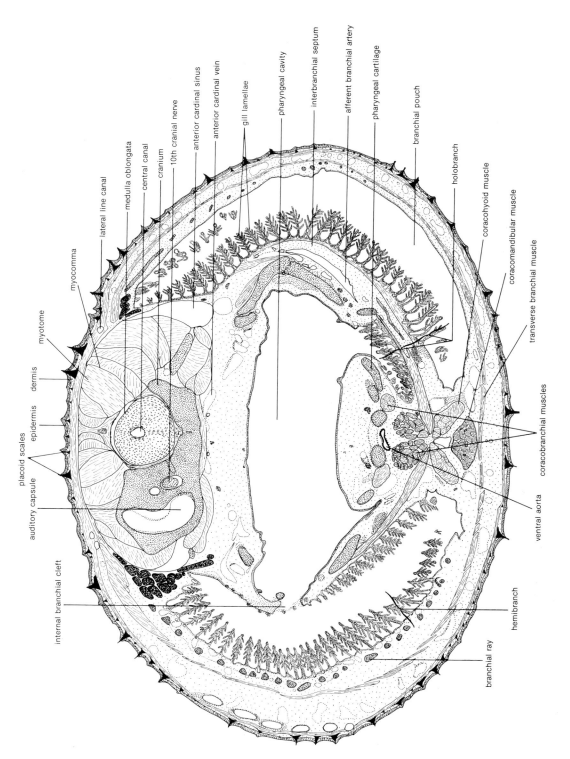

lateral line canal

medulla oblongata

central canal

cranium

10th cranial nerve

anterior cardinal sinus

anterior cardinal vein

gill lamellae

pharyngeal cavity

interbranchial septum

afferent branchial artery

pharyngeal cartilage

branchial pouch

holobranch

coracohyoid muscle

coracomandibular muscle

transverse branchial muscle

myocomma

myotome

dermis

epidermis

placoid scales

auditory capsule

internal branchial cleft

coracobranchial muscles

ventral aorta

hemibranch

branchial ray

DRAWING OF SPECIMEN 35

29

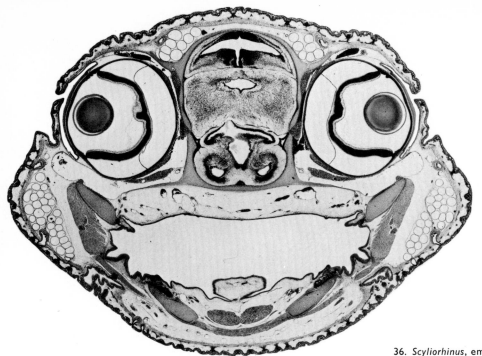

36. *Scyliorhinus*, embryo, eye region TS. (Mag. ×

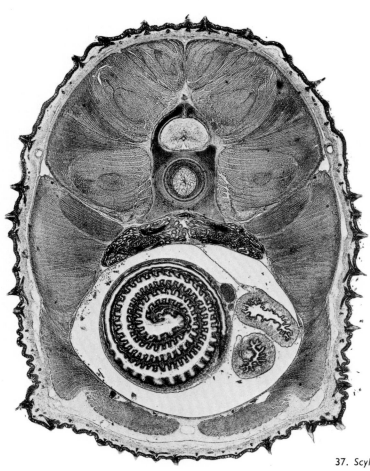

37. *Scyliorhinus*, embryo, intestinal region TS. (Mag. ×

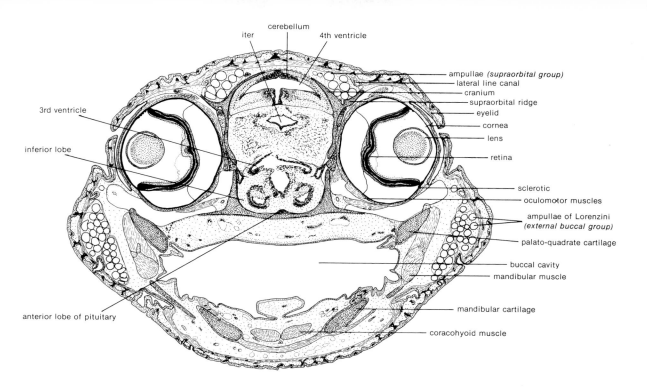

iter
cerebellum
4th ventricle
ampullae (supraorbital group)
lateral line canal
cranium
supraorbital ridge
3rd ventricle
eyelid
cornea
lens
retina
inferior lobe
sclerotic
oculomotor muscles
ampullae of Lorenzini (external buccal group)
palato-quadrate cartilage
buccal cavity
mandibular muscle
anterior lobe of pituitary
mandibular cartilage
coracohyoid muscle

DRAWING OF SPECIMEN 36

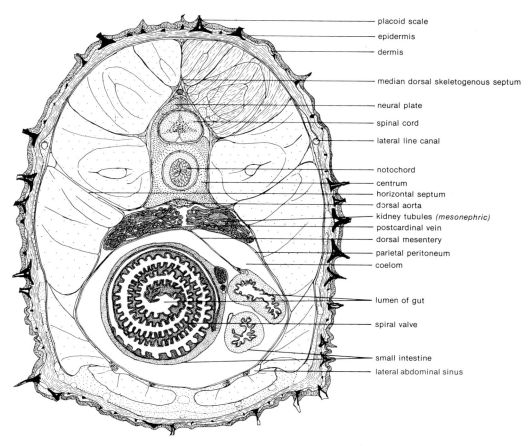

placoid scale
epidermis
dermis
median dorsal skeletogenous septum
neural plate
spinal cord
lateral line canal
notochord
centrum
horizontal septum
dorsal aorta
kidney tubules (mesonephric)
postcardinal vein
dorsal mesentery
parietal peritoneum
coelom
lumen of gut
spiral valve
small intestine
lateral abdominal sinus

DRAWING OF SPECIMEN 37

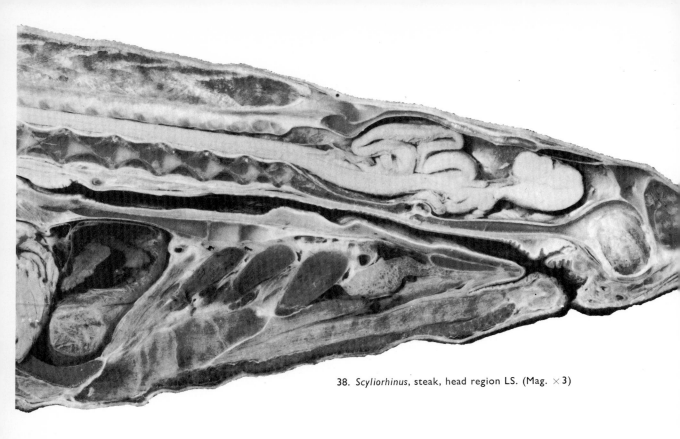

38. *Scyliorhinus*, steak, head region LS. (Mag. ×3)

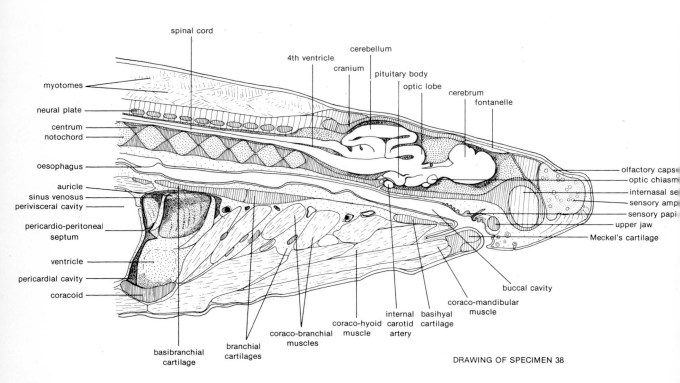

spinal cord

4th ventricle

cerebellum

cranium

pituitary body

optic lobe

cerebrum

fontanelle

myotomes

neural plate

centrum

notochord

oesophagus

auricle

sinus venosus

perivisceral cavity

pericardio-peritoneal septum

ventricle

pericardial cavity

coracoid

olfactory caps[...]

optic chiasm[...]

internasal se[...]

sensory amp[...]

sensory papi[...]

upper jaw

Meckel's cartilage

buccal cavity

coraco-mandibular muscle

internal carotid artery

basihyal cartilage

coraco-hyoid muscle

coraco-branchial muscles

branchial cartilages

basibranchial cartilage

DRAWING OF SPECIMEN 38

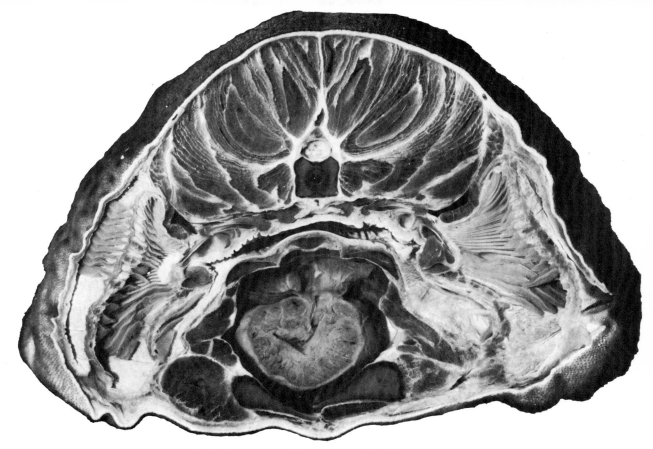

39. *Scyliorhinus*, steak, heart region TS. (Mag. ×2)

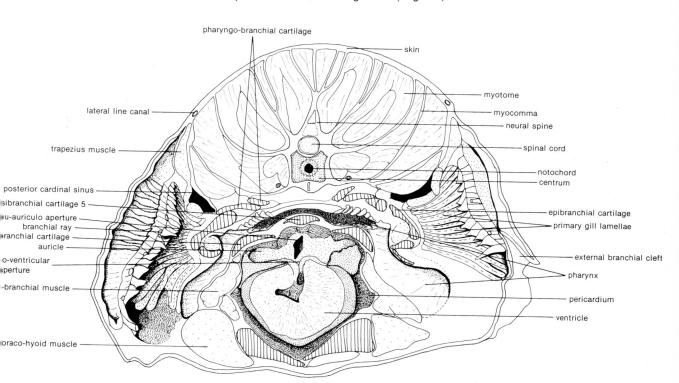

pharyngo-branchial cartilage

skin

lateral line canal

myotome

myocomma

neural spine

trapezius muscle

spinal cord

notochord

centrum

posterior cardinal sinus

sibranchial cartilage 5

u-auriculo aperture

branchial ray

epibranchial cartilage

primary gill lamellae

ranchial cartilage

auricle

o-ventricular
perture

external branchial cleft

pharynx

-branchial muscle

pericardium

ventricle

oraco-hyoid muscle

DRAWING OF SPECIMEN 39

33

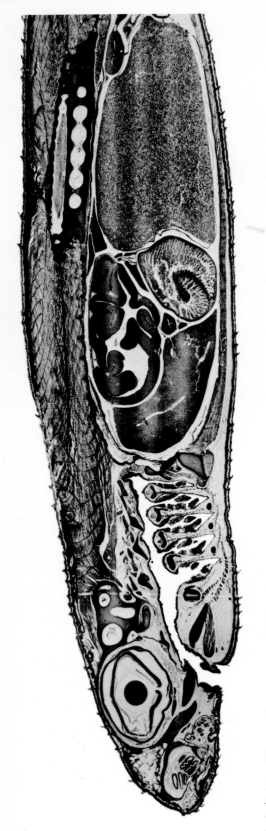

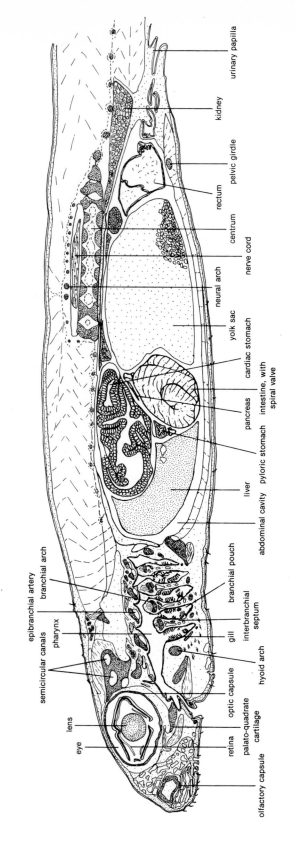

40. *Scyliorhinus*, 100mm embryo, parasagittal LS. (Mag. ×7)

DRAWING BASED ON SPECIMEN 40

urinary papilla
kidney
pelvic girdle
rectum
centrum
nerve cord
neural arch
yolk sac
cardiac stomach
intestine, with spiral valve
pancreas
pyloric stomach
liver
abdominal cavity
branchial pouch
interbranchial septum
gill
hyoid arch
palato-quadrate cartilage
retina
eye
lens
optic capsule
olfactory capsule
semicircular canals
pharynx
epibranchial artery
branchial arch

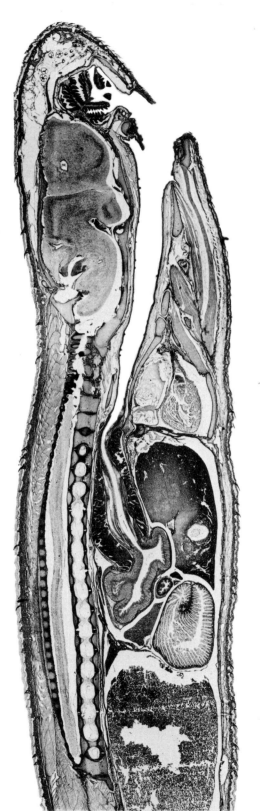

41. *Scyliorhinus*, 100mm embryo, saggittal LS. (Mag. ×7)

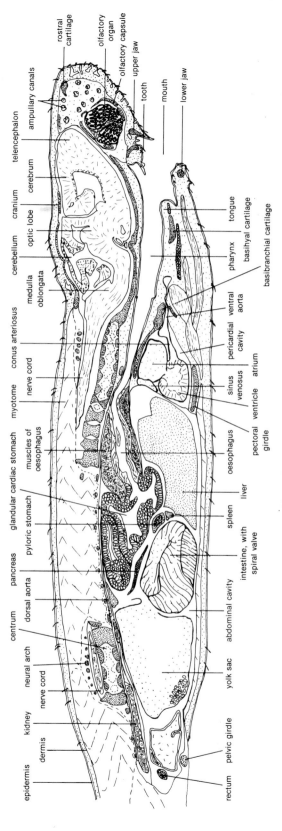

rostral cartilage
olfactory organ
olfactory capsule
upper jaw
tooth
mouth
lower jaw
ampullary canals
telencephalon
cerebrum
cranium
optic lobe
cerebellum
tongue
pharynx
basihyal cartilage
basibranchial cartilage
medulla oblongata
conus arteriosus
ventral aorta
nerve cord
myotome
pericardial cavity
atrium
sinus venosus
ventricle
muscles of oesophagus
glandular cardiac stomach
oesophagus
pectoral girdle
pyloric stomach
liver
pancreas
spleen
dorsal aorta
centrum
intestine, with spiral valve
neural arch
abdominal cavity
nerve cord
kidney
yolk sac
dermis
epidermis
rectum
pelvic girdle

DRAWING BASED ON SPECIMEN 41

35

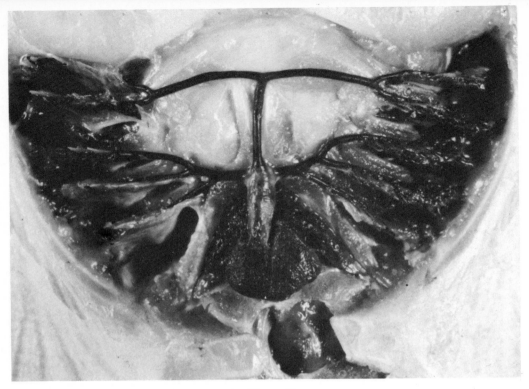

42. *Raja*, dissection of heart and afferent vessels. (Mag. ×3)

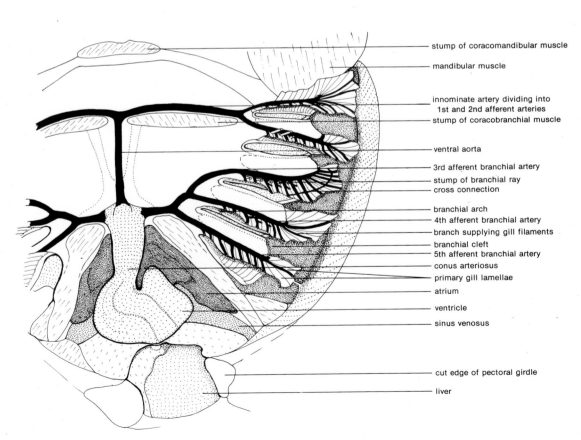

stump of coracomandibular muscle

mandibular muscle

innominate artery dividing into
1st and 2nd afferent arteries

stump of coracobranchial muscle

ventral aorta

3rd afferent branchial artery

stump of branchial ray

cross connection

branchial arch

4th afferent branchial artery

branch supplying gill filaments

branchial cleft

5th afferent branchial artery

conus arteriosus

primary gill lamellae

atrium

ventricle

sinus venosus

cut edge of pectoral girdle

liver

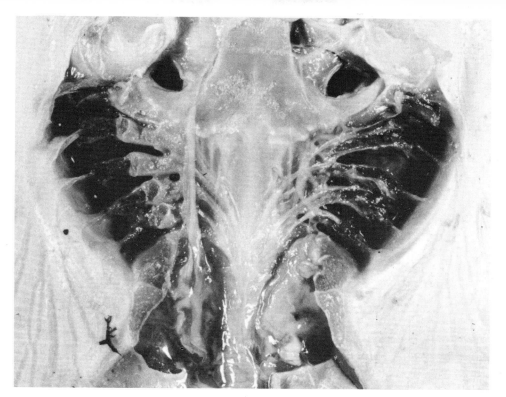

43. *Raja*, dissection of efferent vessels. (Mag. ×2)

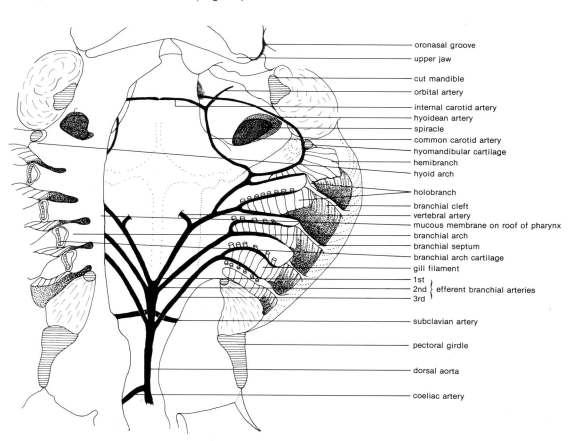

- oronasal groove
- upper jaw
- cut mandible
- orbital artery
- internal carotid artery
- hyoidean artery
- spiracle
- common carotid artery
- hyomandibular cartilage
- hemibranch
- hyoid arch
- holobranch
- branchial cleft
- vertebral artery
- mucous membrane on roof of pharynx
- branchial arch
- branchial septum
- branchial arch cartilage
- gill filament
- 1st
- 2nd } efferent branchial arteries
- 3rd
- subclavian artery
- pectoral girdle
- dorsal aorta
- coeliac artery

DRAWING OF SPECIMEN 43

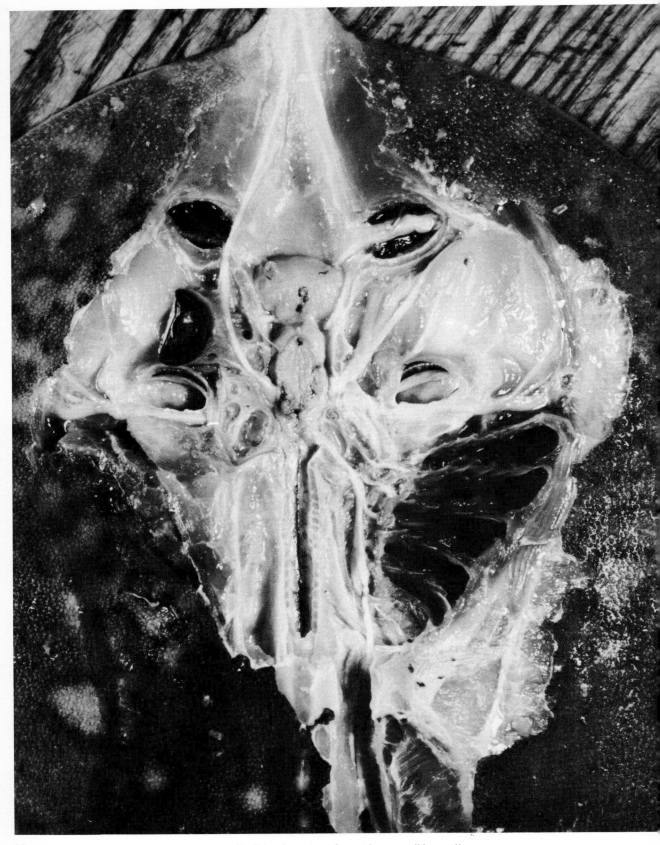

44. *Raja*, dissection of cranial nerves. (Mag. ×1)

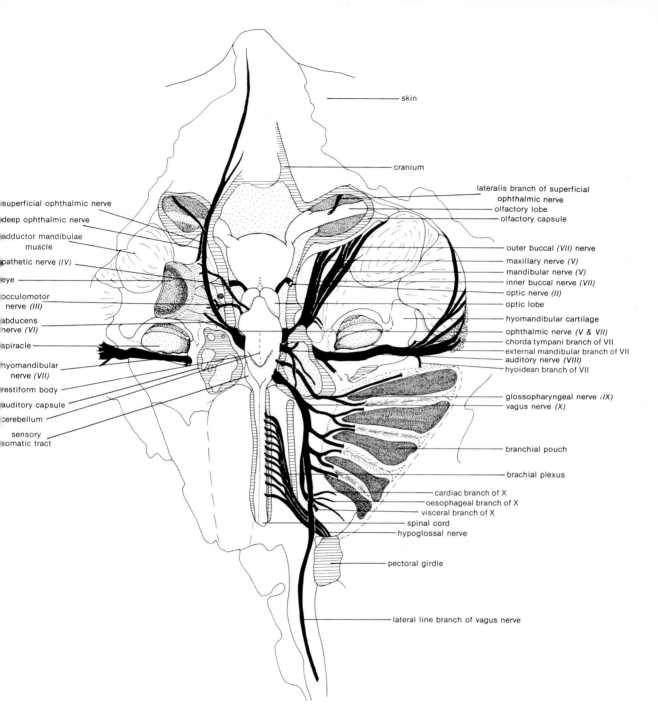

skin

cranium

lateralis branch of superficial
ophthalmic nerve
olfactory lobe
olfactory capsule

superficial ophthalmic nerve

deep ophthalmic nerve

adductor mandibulae
muscle

pathetic nerve (IV)

eye

occulomotor
nerve (III)

abducens
nerve (VI)

spiracle

hyomandibular
nerve (VII)

restiform body

auditory capsule

cerebellum

sensory
somatic tract

outer buccal (VII) nerve
maxillary nerve (V)
mandibular nerve (V)
inner buccal nerve (VII)
optic nerve (II)
optic lobe
hyomandibular cartilage
ophthalmic nerve (V & VII)
chorda tympani branch of VII
external mandibular branch of VII
auditory nerve (VIII)
hyoidean branch of VII

glossopharyngeal nerve (IX)
vagus nerve (X)

branchial pouch

brachial plexus

cardiac branch of X
oesophageal branch of X
visceral branch of X
spinal cord
hypoglossal nerve

pectoral girdle

lateral line branch of vagus nerve

DRAWING OF SPECIMEN 44

39

45. *Gadus merlangus* general dissection, (Mag. ×1)

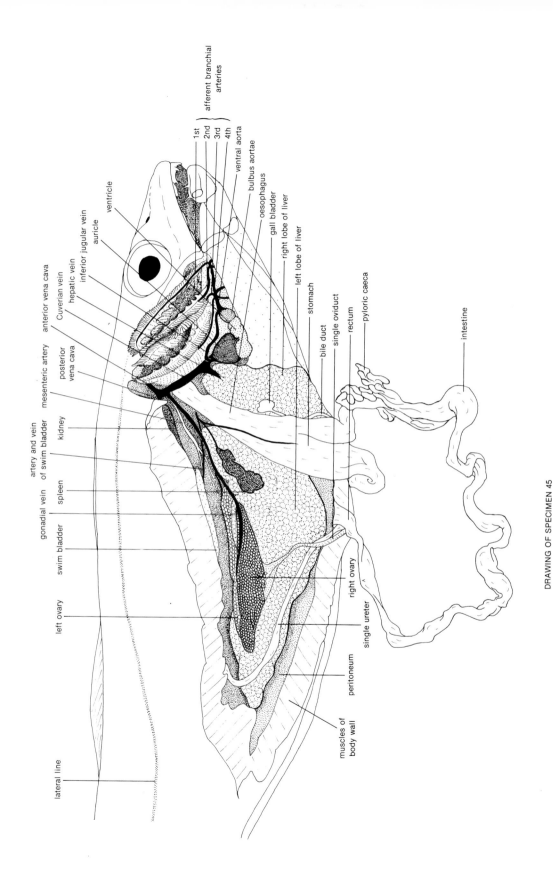

DRAWING OF SPECIMEN 45

afferent branchial arteries

1st
2nd
3rd
4th

ventral aorta

bulbus aortae

oesophagus

gall bladder

right lobe of liver

left lobe of liver

stomach

bile duct

single oviduct

rectum

pyloric caeca

intestine

ventricle

inferior jugular vein

auricle

hepatic vein

Cuverian vein

anterior vena cava

posterior
vena cava

mesenteric artery

kidney

artery and vein
of swim bladder

spleen

gonadial vein

swim bladder

left ovary

lateral line

right ovary

single ureter

peritoneum

muscles of
body wall

41

46. *Gadus morrhuae*, dissection of cranial nerves. (Mag. ×1)

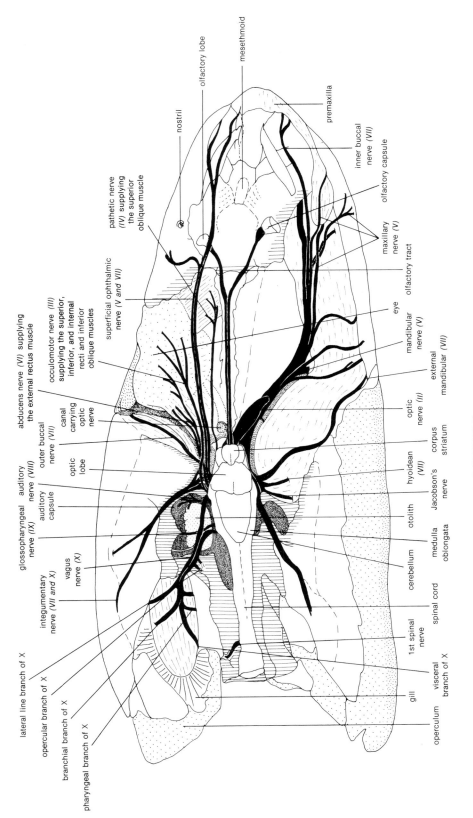

nostril

olfactory lobe

mesethmoid

premaxilla

inner buccal nerve (VII)

olfactory capsule

maxillary nerve (V)

olfactory tract

pathetic nerve (IV) supplying the superior oblique muscle

superficial ophthalmic nerve (V and VII)

eye

mandibular nerve (V)

external mandibular (VII)

abducens nerve (VI) supplying the external rectus muscle

occulomotor nerve (III) supplying the superior, inferior, and internal recti and inferior oblique muscles

canal carrying optic nerve

optic nerve (II)

corpus striatum

auditory nerve (VIII)

outer buccal nerve (VII)

optic lobe

hyoidean (VII)

glossopharyngeal nerve (IX)

auditory capsule

Jacobson's nerve

otolith

vagus nerve (X)

cerebellum

medulla oblongata

integumentary nerve (VII and X)

spinal cord

1st spinal nerve

lateral line branch of X

opercular branch of X

branchial branch of X

pharyngeal branch of X

gill

visceral branch of X

operculum

DRAWING OF SPECIMEN 46

43

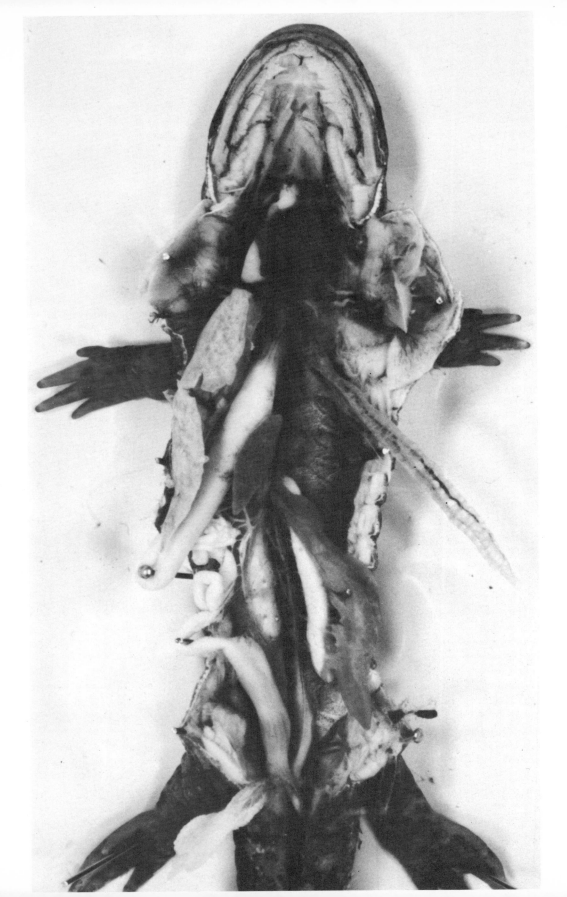

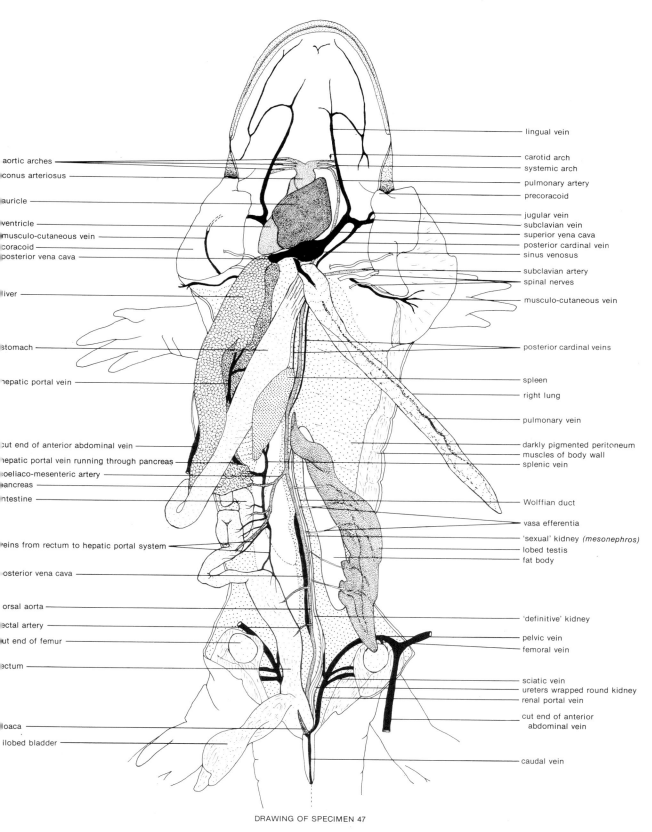

lingual vein

aortic arches

conus arteriosus

auricle

ventricle

musculo-cutaneous vein

coracoid

posterior vena cava

liver

stomach

hepatic portal vein

cut end of anterior abdominal vein

hepatic portal vein running through pancreas

coeliaco-mesenteric artery

pancreas

intestine

veins from rectum to hepatic portal system

posterior vena cava

dorsal aorta

rectal artery

cut end of femur

rectum

cloaca

bilobed bladder

carotid arch

systemic arch

pulmonary artery

precoracoid

jugular vein

subclavian vein

superior vena cava

posterior cardinal vein

sinus venosus

subclavian artery

spinal nerves

musculo-cutaneous vein

posterior cardinal veins

spleen

right lung

pulmonary vein

darkly pigmented peritoneum

muscles of body wall

splenic vein

Wolffian duct

vasa efferentia

'sexual' kidney *(mesonephros)*

lobed testis

fat body

'definitive' kidney

pelvic vein

femoral vein

sciatic vein

ureters wrapped round kidney

renal portal vein

cut end of anterior
abdominal vein

caudal vein

DRAWING OF SPECIMEN 47

Salamandra, general dissection. (Mag. ×2)

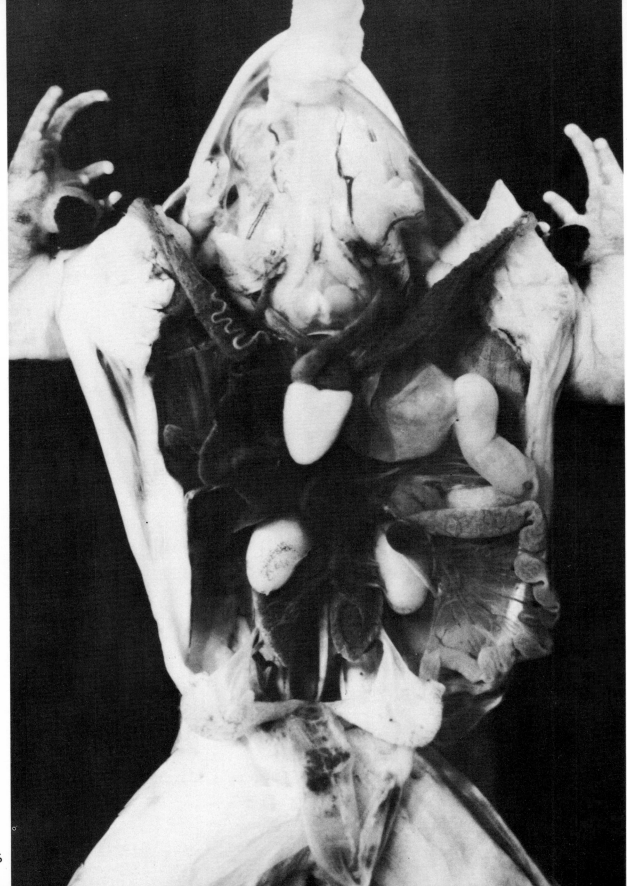

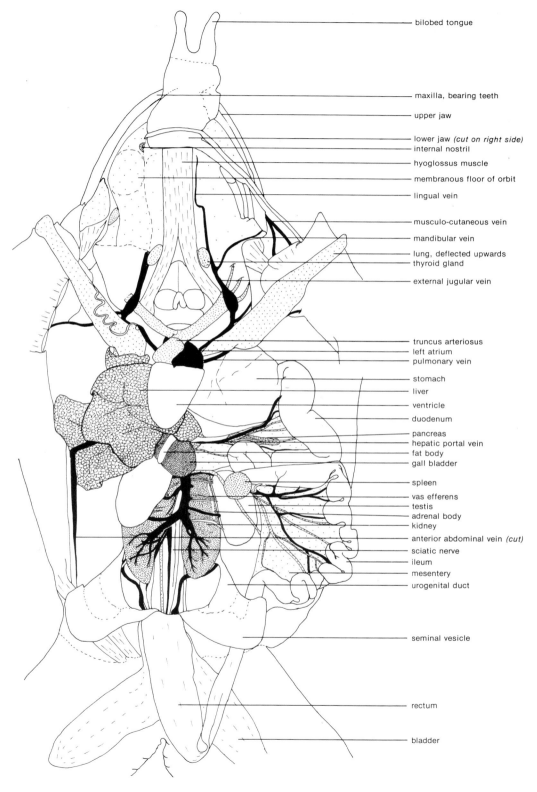

bilobed tongue

maxilla, bearing teeth

upper jaw

lower jaw (cut on right side)
internal nostril
hyoglossus muscle
membranous floor of orbit
lingual vein

musculo-cutaneous vein

mandibular vein

lung, deflected upwards
thyroid gland

external jugular vein

truncus arteriosus
left atrium
pulmonary vein
stomach
liver
ventricle
duodenum
pancreas
hepatic portal vein
fat body
gall bladder

spleen
vas efferens
testis
adrenal body
kidney
anterior abdominal vein (cut)
sciatic nerve
ileum
mesentery
urogenital duct

seminal vesicle

rectum

bladder

DRAWING OF SPECIMEN 48

48. *Rana*, general dissection. (Mag. ×3)

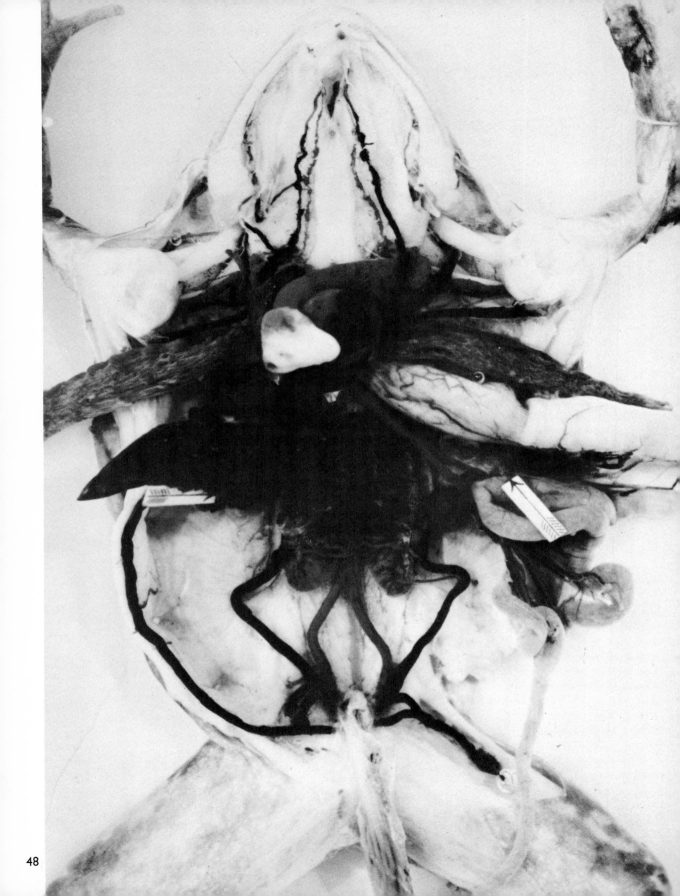

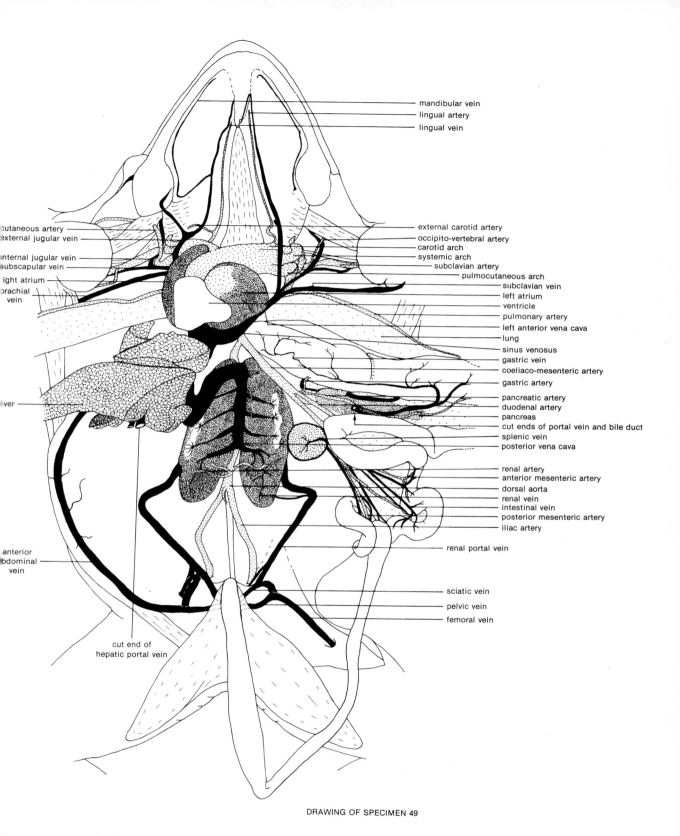

mandibular vein
lingual artery
lingual vein

external carotid artery
occipito-vertebral artery
carotid arch
systemic arch
subclavian artery
pulmocutaneous arch
subclavian vein
left atrium
ventricle
pulmonary artery
left anterior vena cava
lung
sinus venosus
gastric vein
coeliaco-mesenteric artery
gastric artery
pancreatic artery
duodenal artery
pancreas
cut ends of portal vein and bile duct
splenic vein
posterior vena cava
renal artery
anterior mesenteric artery
dorsal aorta
renal vein
intestinal vein
posterior mesenteric artery
iliac artery
renal portal vein
sciatic vein
pelvic vein
femoral vein

cutaneous artery
external jugular vein
internal jugular vein
subscapular vein
right atrium
brachial vein
liver
anterior abdominal vein
cut end of hepatic portal vein

DRAWING OF SPECIMEN 49

Rana, dissection of blood vascular system. (Mag. × 3)

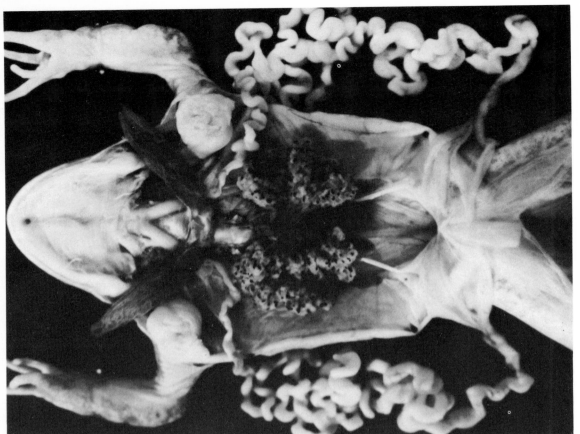

51. *Rana*, dissection of female urogenital system. (Mag. ×2)

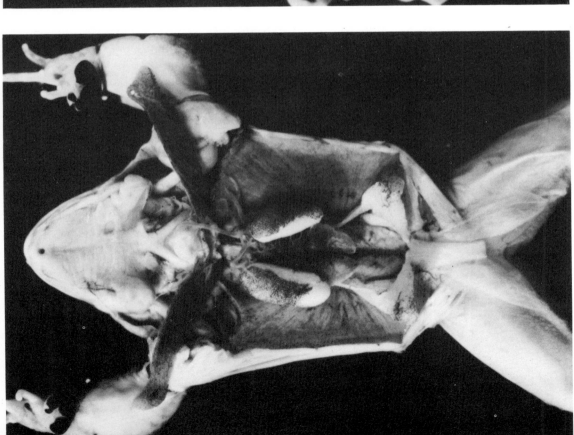

50. *Rana*, dissection of male urogenital system. (Mag. ×2)

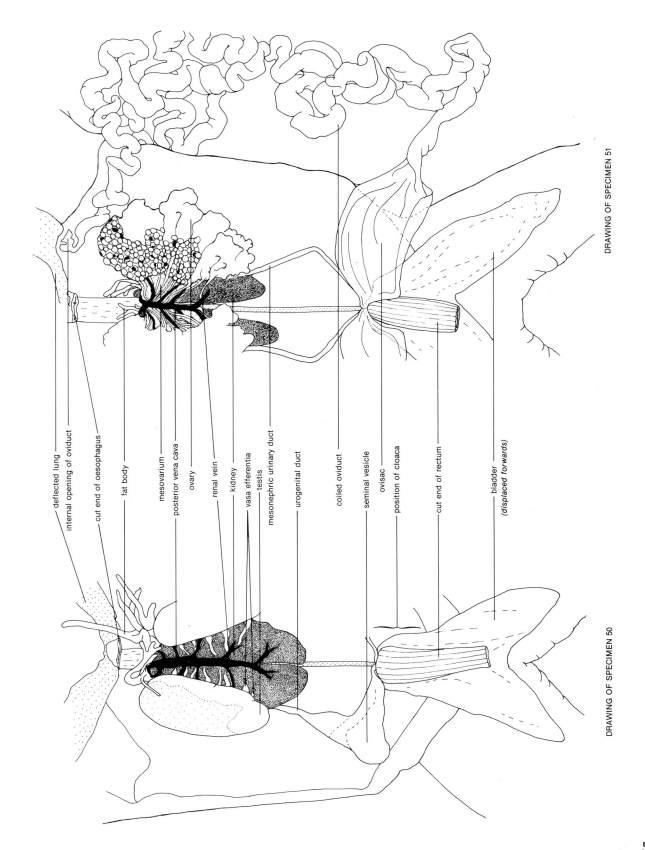

deflected lung

internal opening of oviduct

cut end of oesophagus

fat body

mesovarium

posterior vena cava

ovary

renal vein

kidney

vasa efferentia

testis

mesonephric urinary duct

urogenital duct

coiled oviduct

seminal vesicle

ovisac

position of cloaca

cut end of rectum

bladder
(displaced forwards)

DRAWING OF SPECIMEN 51

DRAWING OF SPECIMEN 50

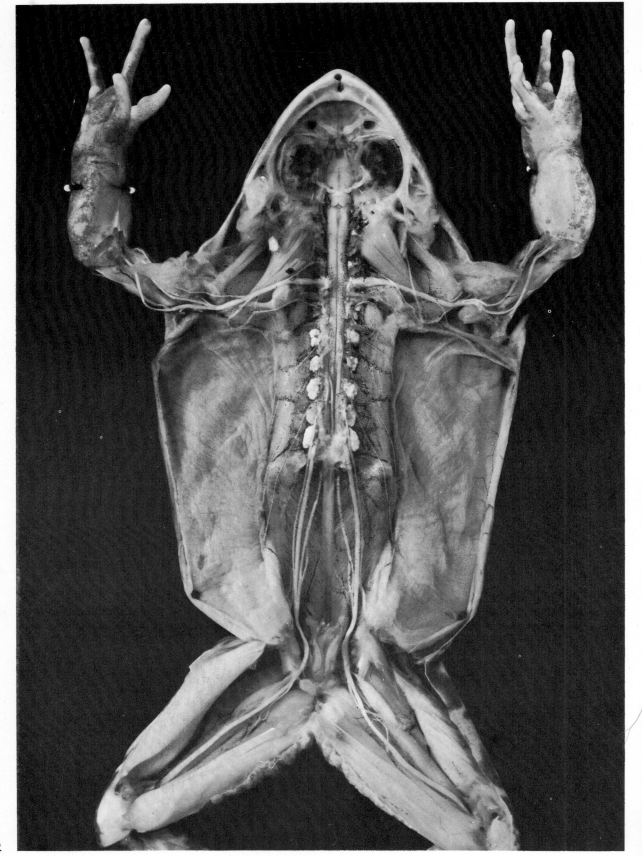

52

52. *Rana*, dissection of spinal nerves. (Mag. ×2)

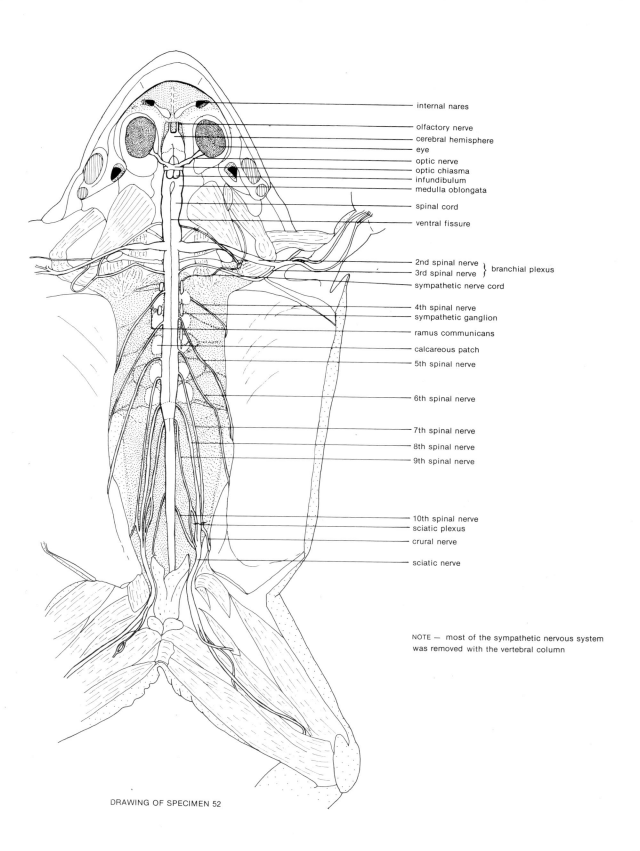

internal nares
olfactory nerve
cerebral hemisphere
eye
optic nerve
optic chiasma
infundibulum
medulla oblongata

spinal cord

ventral fissure

2nd spinal nerve } branchial plexus
3rd spinal nerve
sympathetic nerve cord

4th spinal nerve
sympathetic ganglion

ramus communicans

calcareous patch

5th spinal nerve

6th spinal nerve

7th spinal nerve

8th spinal nerve

9th spinal nerve

10th spinal nerve
sciatic plexus

crural nerve

sciatic nerve

NOTE — most of the sympathetic nervous system
was removed with the vertebral column

DRAWING OF SPECIMEN 52

53

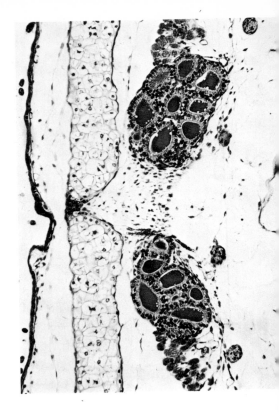

53. *Rana*, entire steak LS. (Mag. ×2)

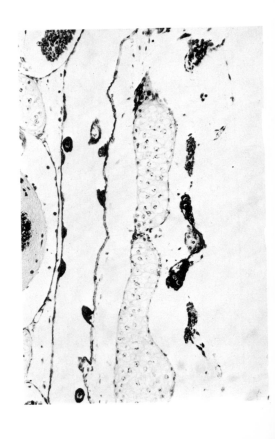

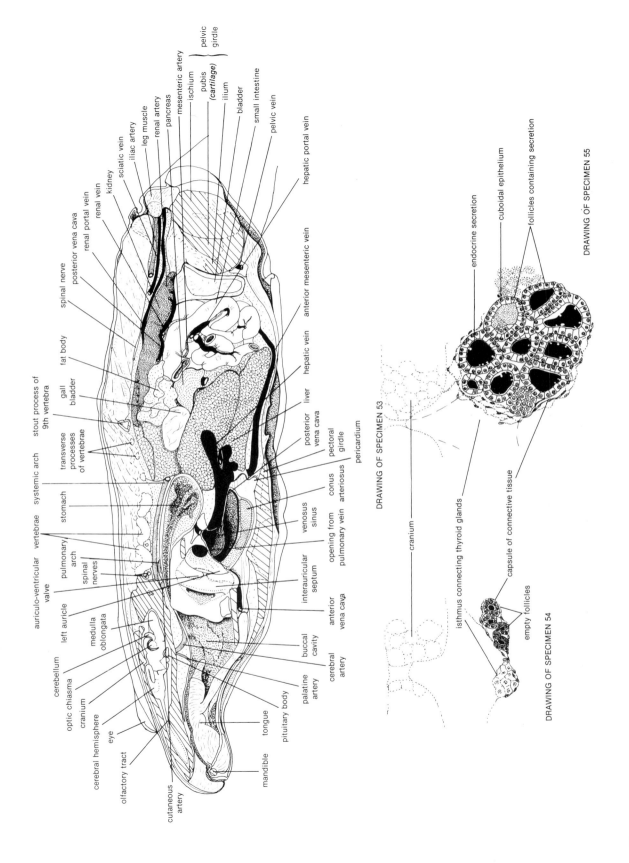

cutaneous artery
olfactory tract
mandible
eye
cranium
optic chiasma
cerebral hemisphere
cerebellum
left auricle
medulla oblongata
auriculo-ventricular valve
vertebrae
systemic arch
stout process of 9th vertebra
transverse processes of vertebrae
pulmonary arch
spinal nerves
stomach
fat body
gall bladder
spinal nerve
posterior vena cava
renal portal vein
renal vein
kidney
sciatic vein
iliac artery
leg muscle
renal artery
pancreas
mesenteric artery
ischium
pubis (cartilage)
pelvic girdle
ilium
bladder
small intestine
pelvic vein
hepatic portal vein
anterior mesenteric vein
hepatic vein
liver
posterior vena cava
pectoral girdle
pericardium
venosus sinus
conus arteriosus
opening from pulmonary vein
interauricular septum
anterior vena cava
buccal cavity
cerebral artery
palatine artery
pituitary body
tongue

DRAWING OF SPECIMEN 53

cranium
isthmus connecting thyroid glands
capsule of connective tissue
empty follicles

DRAWING OF SPECIMEN 54

endocrine secretion
cuboidal epithelium
follicles containing secretion

DRAWING OF SPECIMEN 55

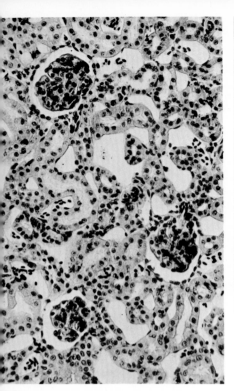

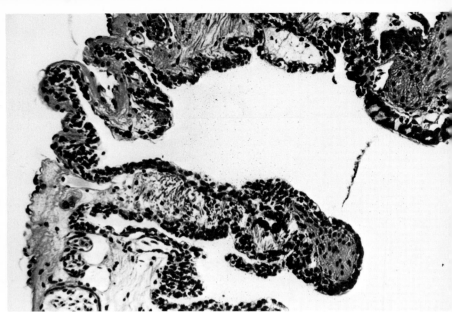

56. *Rana*, kidney TS. (Mag. ×100)

58. *Rana*, lung TS. (Mag. ×10

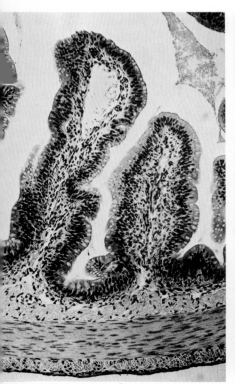

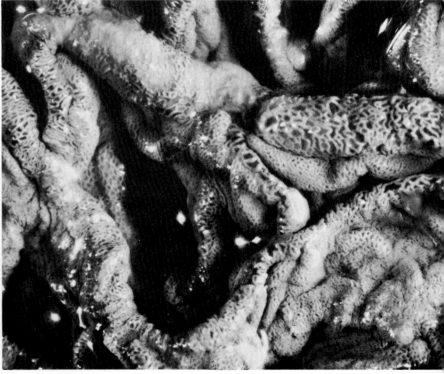

57. *Rana*, intestine TS. (Mag. ×100)

59. Bufo, lung injected, inner surface view by reflected light. (Mag. ×50)

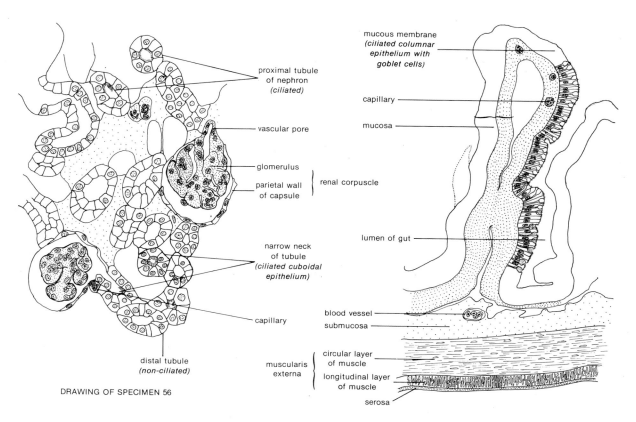

proximal tubule
of nephron
(ciliated)

vascular pore

glomerulus

parietal wall
of capsule

renal corpuscle

narrow neck
of tubule
*(ciliated cuboidal
epithelium)*

capillary

distal tubule
(non-ciliated)

muscularis
externa

DRAWING OF SPECIMEN 56

mucous membrane
*(ciliated columnar
epithelium with
goblet cells)*

capillary

mucosa

lumen of gut

blood vessel

submucosa

circular layer
of muscle

longitudinal layer
of muscle

serosa

DRAWING OF SPECIMEN 57

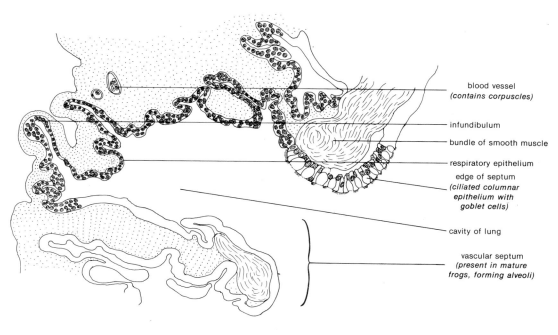

blood vessel
(contains corpuscles)

infundibulum

bundle of smooth muscle

respiratory epithelium

edge of septum
*(ciliated columnar
epithelium with
goblet cells)*

cavity of lung

vascular septum
*(present in mature
frogs, forming alveoli)*

DRAWING OF SPECIMEN 58

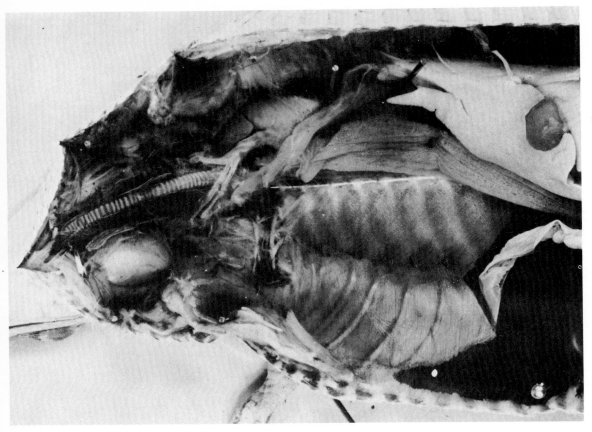

61. *Lacerta*, dissection of heart and arterial system. (Mag. ×7)

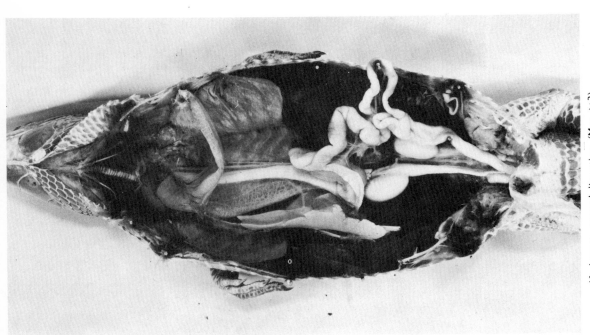

60. *Lacerta*, general dissection. (Mag. ×3)

DRAWING OF SPECIMEN 61

ductus caroticus
left systemic arch
right pulmonary arch
right systemic arch
left auricle
left pulmonary arch

left internal jugular vein
left carotid arch
right internal carotid artery
right internal jugular vein
(partly removed to display arteries)
right external carotid artery
right systemic artery
right superior vena cava
right auricle
(ventral wall removed)
cut end of subclavian vein
subclavian artery
brachial nerve
position of ventricle
azygos vein
artery to body wall
posterior vena cava
dorsal aorta
pulmonary artery
right lung
pulmonary vein
liver
gall bladder
anterior abdominal vein
posterior vena cava

basilingual plate of hyoid
posterior cornu of hyoid
trachea
right external jugular vein
right auricle
left auricle
ventricle
lightly pigmented peritoneum
posterior vena cava
lungs
oesophagus
pulmonary artery
dorsal aorta
liver
darkly pigmented peritoneum
stomach
pancreas
hepatic portal vein entering liver
hepatic portal vein
duodenum
spleen
coeliaco-mesenteric artery
intestine
cut end of anterior abdominal vein
testis
epididymis
right posterior vena cava
left posterior vena cava
pelvic vein
cut end of femur
rectum
kidney
cloaca
bilobed bladder

DRAWING OF SPECIMEN 60

59

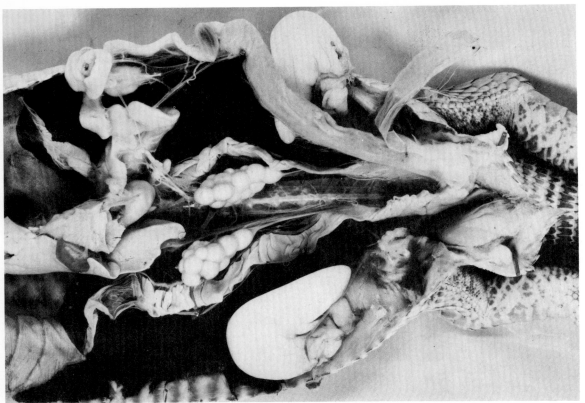

63. *Lacerta*, dissection of female urogenital system. (Mag. ×7)

62. *Lacerta*, dissection of male urogenital system. (Mag. ×7)

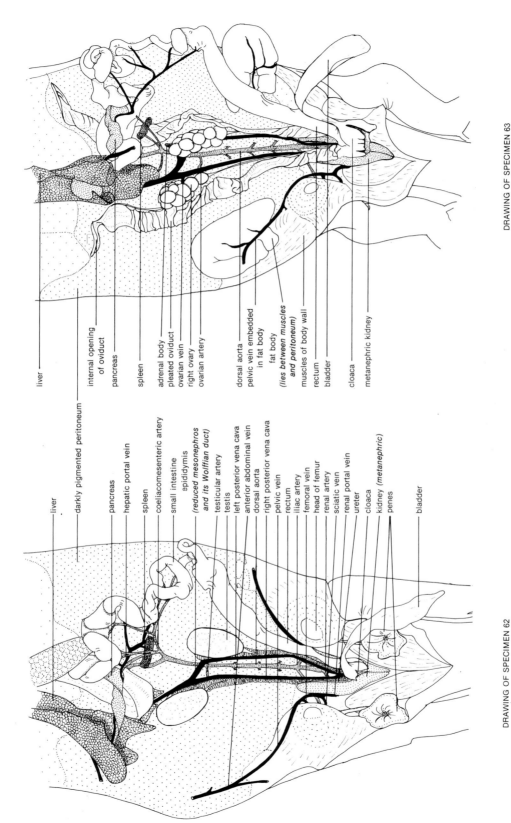

DRAWING OF SPECIMEN 63

liver

internal opening
of oviduct

pancreas

spleen

adrenal body
pleated oviduct
ovarian vein
right ovary
ovarian artery

dorsal aorta
pelvic vein embedded
in fat body
fat body
(lies between muscles
and peritoneum)
muscles of body wall
rectum
bladder

cloaca

metanephric kidney

DRAWING OF SPECIMEN 62

liver

darkly pigmented peritoneum

pancreas
hepatic portal vein

spleen
coeliacomesenteric artery

small intestine
epididymis
(reduced mesonephros
and its Wolffian duct)
testicular artery
testis
left posterior vena cava
anterior abdominal vein
dorsal aorta
right posterior vena cava
pelvic vein
rectum
iliac artery
femoral vein
head of femur
renal artery
sciatic vein
renal portal vein
ureter
cloaca
kidney (metanephric)
penes

bladder

61

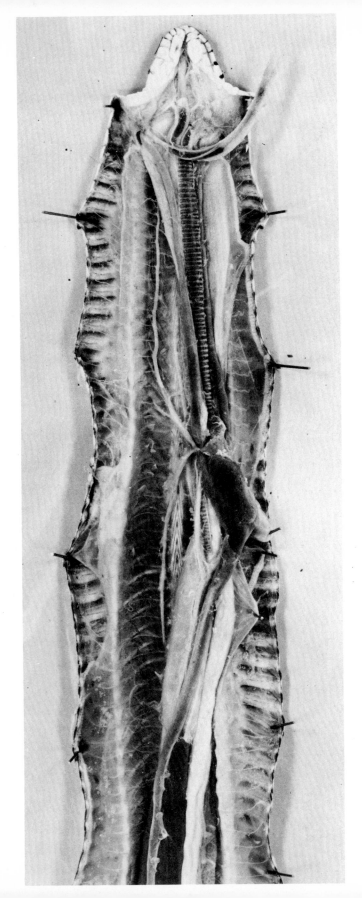

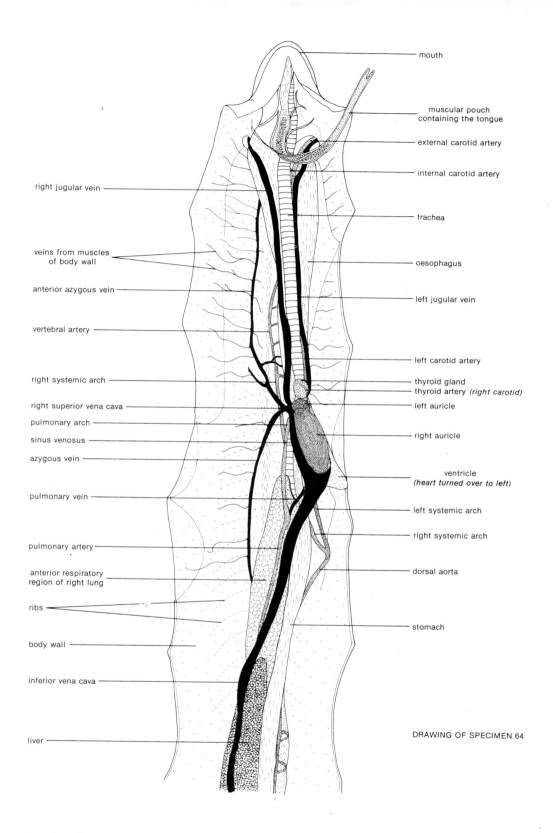

mouth

muscular pouch
containing the tongue

external carotid artery

internal carotid artery

right jugular vein

trachea

veins from muscles
of body wall

oesophagus

anterior azygous vein

left jugular vein

vertebral artery

left carotid artery

right systemic arch

thyroid gland
thyroid artery *(right carotid)*

right superior vena cava

left auricle

pulmonary arch

sinus venosus

right auricle

azygous vein

ventricle
(heart turned over to left)

pulmonary vein

left systemic arch

right systemic arch

pulmonary artery

anterior respiratory
region of right lung

dorsal aorta

ribs

body wall

stomach

inferior vena cava

liver

DRAWING OF SPECIMEN 64

Tropidonotus, dissection of anterior third. (Mag. ×2)

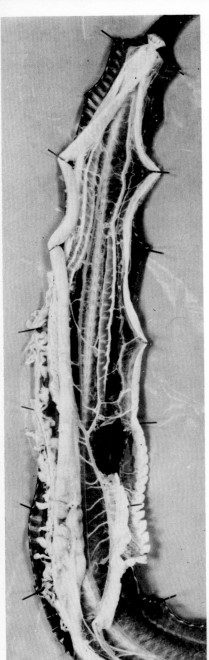

66. *Tropidonotus*, dissection of posterior third. (Mag. ×1·5)

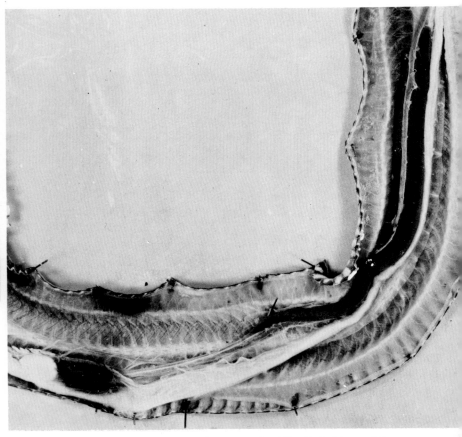

65. *Tropidonotus*, dissection of middle third. (Mag. ×1·5)

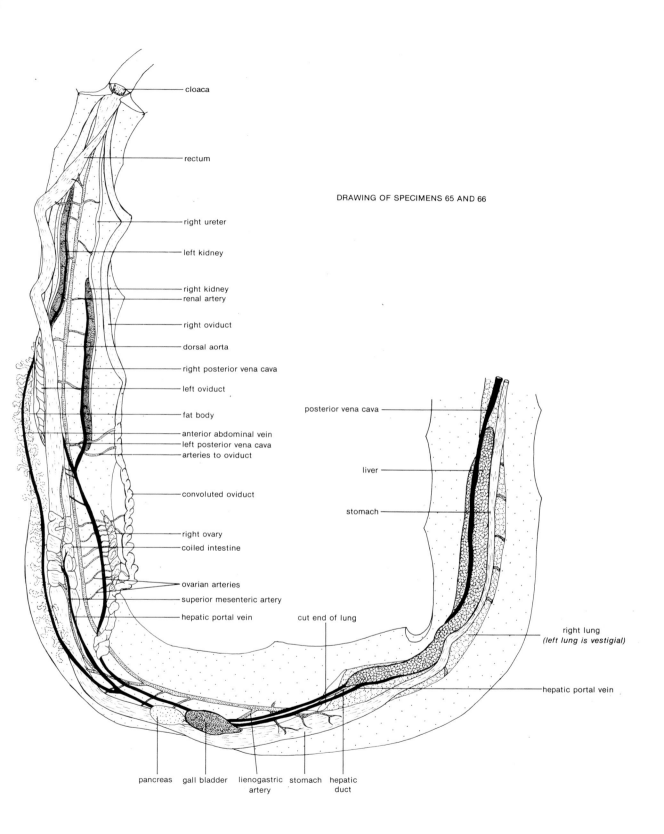

cloaca

rectum

DRAWING OF SPECIMENS 65 AND 66

right ureter

left kidney

right kidney
renal artery

right oviduct

dorsal aorta

right posterior vena cava

left oviduct

fat body

posterior vena cava

anterior abdominal vein
left posterior vena cava
arteries to oviduct

liver

convoluted oviduct

stomach

right ovary
coiled intestine

ovarian arteries
superior mesenteric artery

hepatic portal vein

cut end of lung

right lung
(left lung is vestigial)

hepatic portal vein

pancreas   gall bladder   lienogastric   stomach   hepatic
                              artery   duct

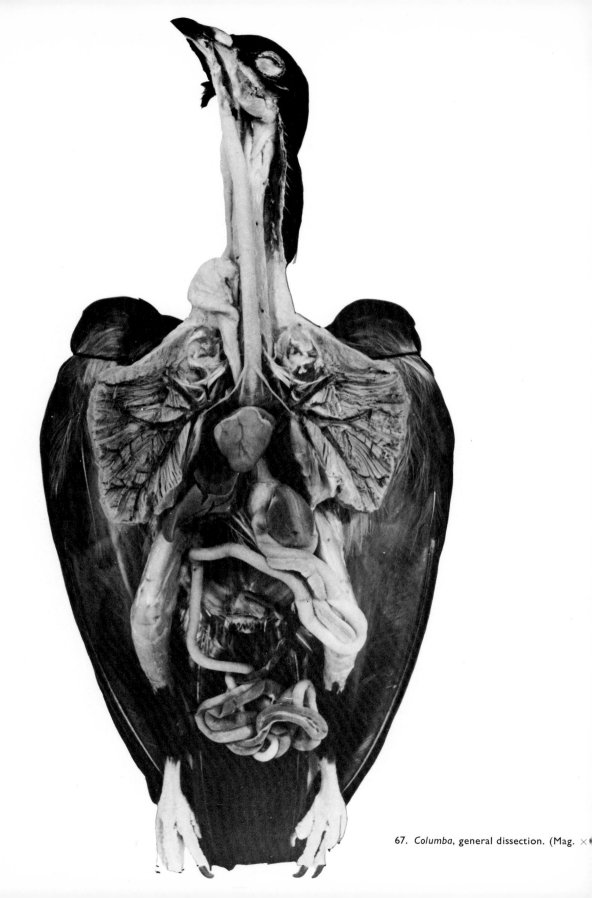

67. *Columba*, general dissection. (Mag. ×

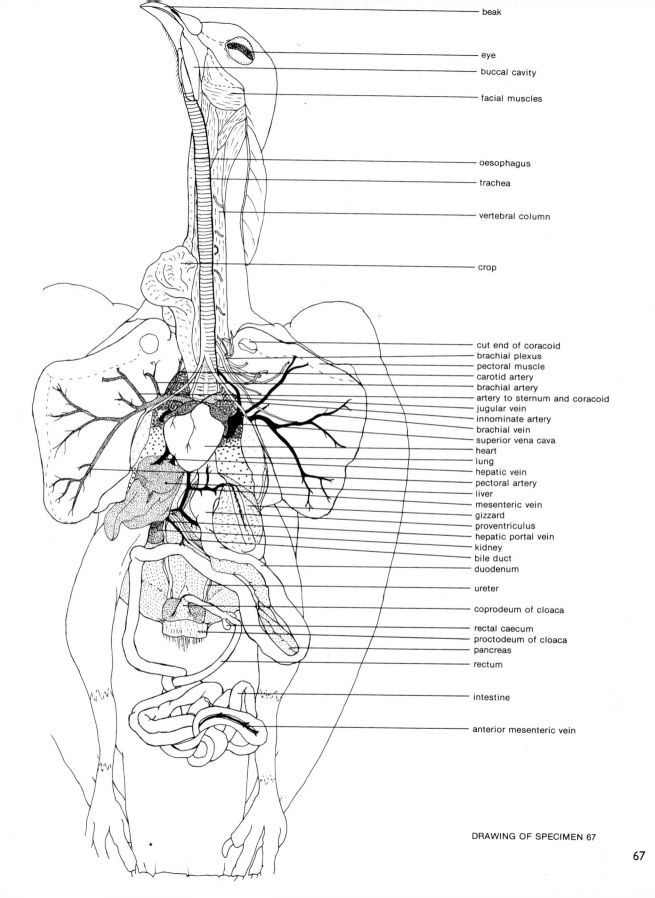

beak

eye

buccal cavity

facial muscles

oesophagus

trachea

vertebral column

crop

cut end of coracoid
brachial plexus
pectoral muscle
carotid artery
brachial artery
artery to sternum and coracoid
jugular vein
innominate artery
brachial vein
superior vena cava
heart
lung
hepatic vein
pectoral artery
liver
mesenteric vein
gizzard
proventriculus
hepatic portal vein
kidney
bile duct
duodenum

ureter

coprodeum of cloaca

rectal caecum
proctodeum of cloaca
pancreas
rectum

intestine

anterior mesenteric vein

DRAWING OF SPECIMEN 67

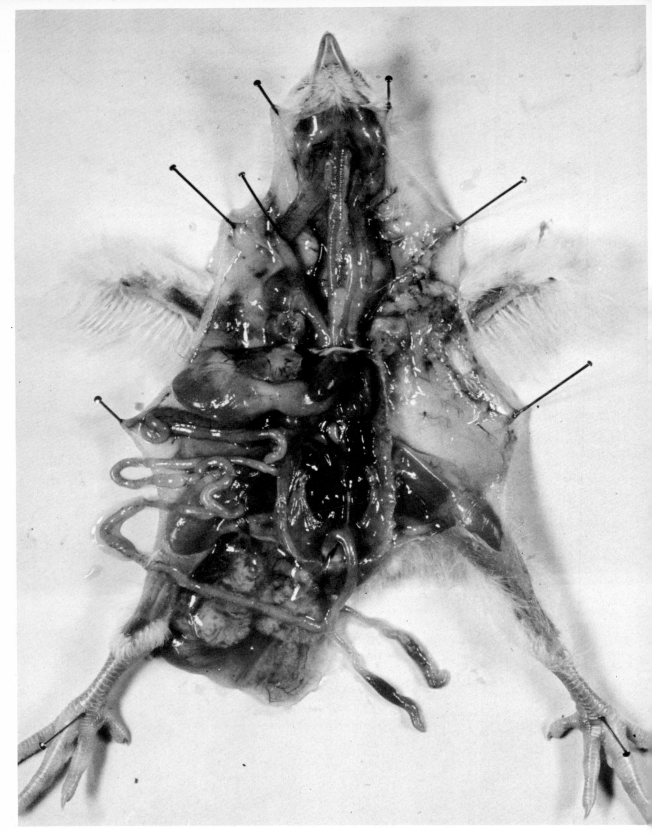

68. *Gallus*, dissection of day-old chick. (Mag. ×2)

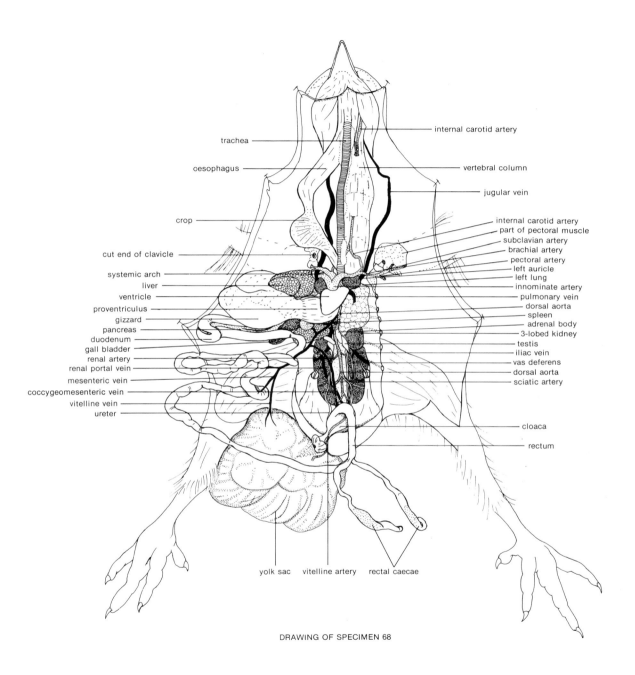

trachea

oesophagus

crop

cut end of clavicle

systemic arch
liver
ventricle
proventriculus
gizzard
pancreas
duodenum
gall bladder
renal artery
renal portal vein
mesenteric vein
coccygeomesenteric vein
vitelline vein
ureter

internal carotid artery

vertebral column

jugular vein

internal carotid artery
part of pectoral muscle
subclavian artery
brachial artery
pectoral artery
left auricle
left lung
innominate artery
pulmonary vein
dorsal aorta
spleen
adrenal body
3-lobed kidney
testis
iliac vein
vas deferens
dorsal aorta
sciatic artery

cloaca

rectum

yolk sac   vitelline artery   rectal caecae

DRAWING OF SPECIMEN 68

69

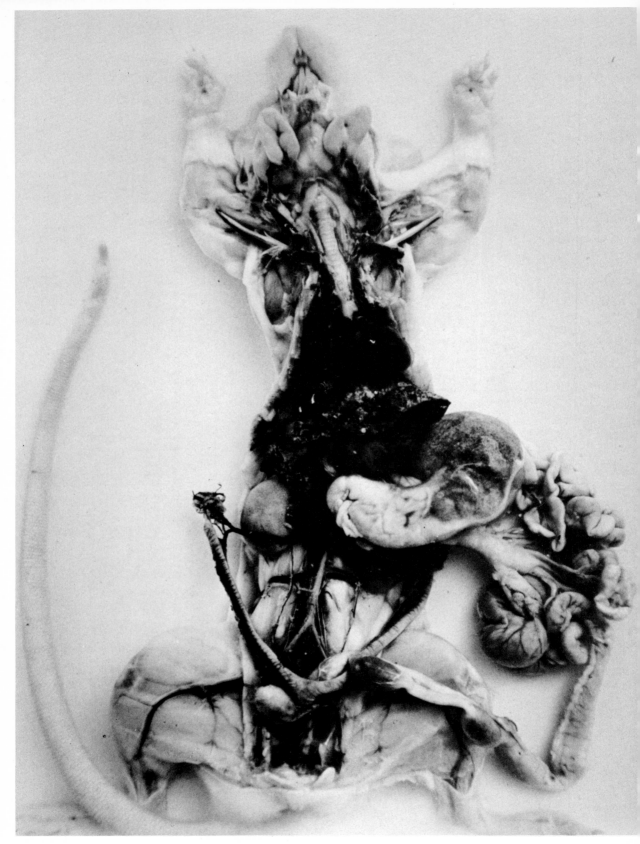

69. *Rattus*, general dissection. (Mag. ×1·2)

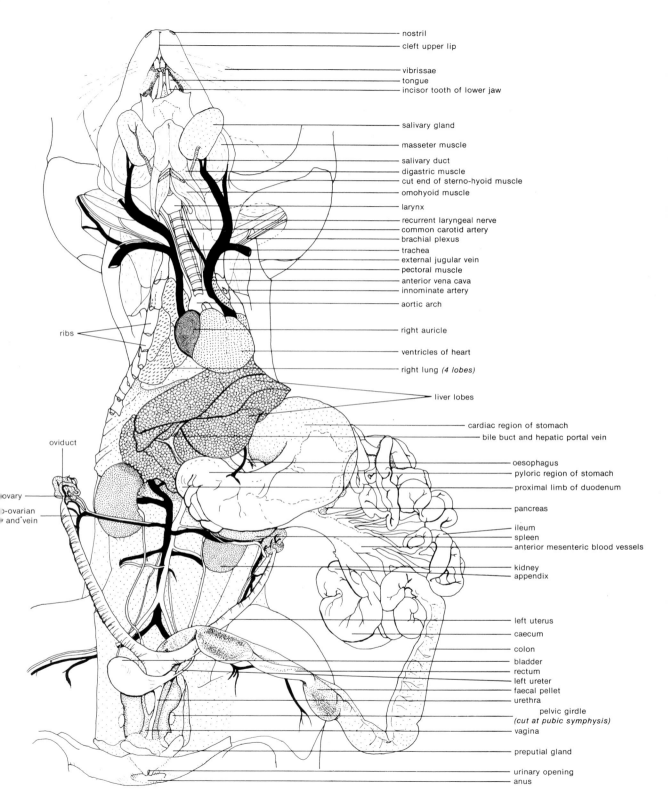

nostril
cleft upper lip

vibrissae
tongue
incisor tooth of lower jaw

salivary gland

masseter muscle

salivary duct
digastric muscle
cut end of sterno-hyoid muscle
omohyoid muscle

larynx

recurrent laryngeal nerve
common carotid artery
brachial plexus
trachea
external jugular vein
pectoral muscle
anterior vena cava
innominate artery

aortic arch

right auricle

ventricles of heart

right lung (4 lobes)

liver lobes

cardiac region of stomach
bile buct and hepatic portal vein

oesophagus
pyloric region of stomach
proximal limb of duodenum

pancreas

ileum
spleen
anterior mesenteric blood vessels

kidney
appendix

left uterus
caecum
colon
bladder
rectum
left ureter
faecal pellet
urethra
pelvic girdle
(cut at pubic symphysis)
vagina

preputial gland

urinary opening
anus

ribs

oviduct

ovary

o-ovarian
 and vein

DRAWING OF SPECIMEN 69

71

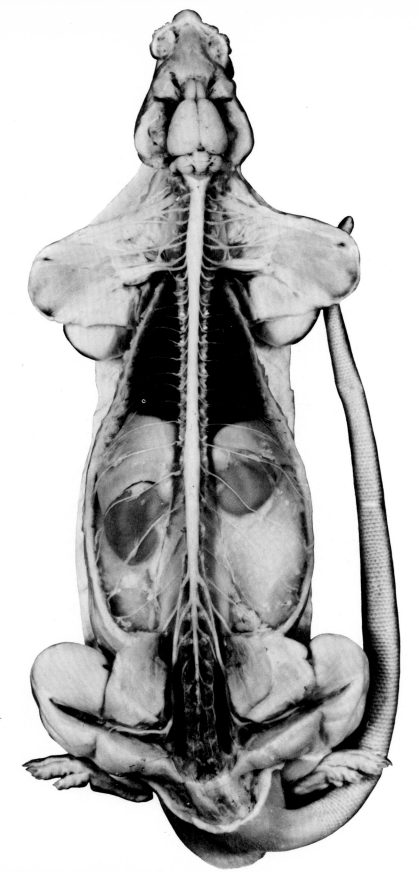

70. *Rattus*, dissection of brain and spinal nerves. (Mag. ×1·1)

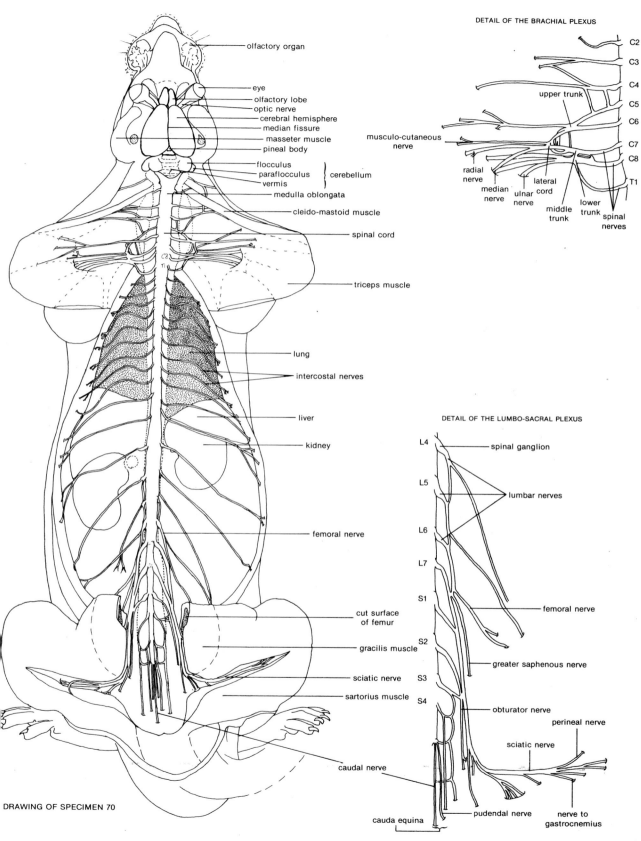

DETAIL OF THE BRACHIAL PLEXUS

olfactory organ

eye
olfactory lobe
optic nerve
cerebral hemisphere
median fissure
masseter muscle
pineal body

flocculus
paraflocculus } cerebellum
vermis
medulla oblongata

cleido-mastoid muscle

spinal cord

triceps muscle

lung

intercostal nerves

liver

kidney

femoral nerve

cut surface
of femur

gracilis muscle

sciatic nerve

sartorius muscle

caudal nerve

DRAWING OF SPECIMEN 70

C2
C3
C4
C5
C6
C7
C8
T1

upper trunk

musculo-cutaneous
nerve

radial
nerve

median
nerve

ulnar
nerve

lateral
cord

middle
trunk

lower
trunk

spinal
nerves

DETAIL OF THE LUMBO-SACRAL PLEXUS

L4   spinal ganglion

L5   lumbar nerves

L6

L7

S1

S2   femoral nerve

S3   greater saphenous nerve

S4   obturator nerve

perineal nerve

sciatic nerve

cauda equina   pudendal nerve   nerve to
gastrocnemius

73

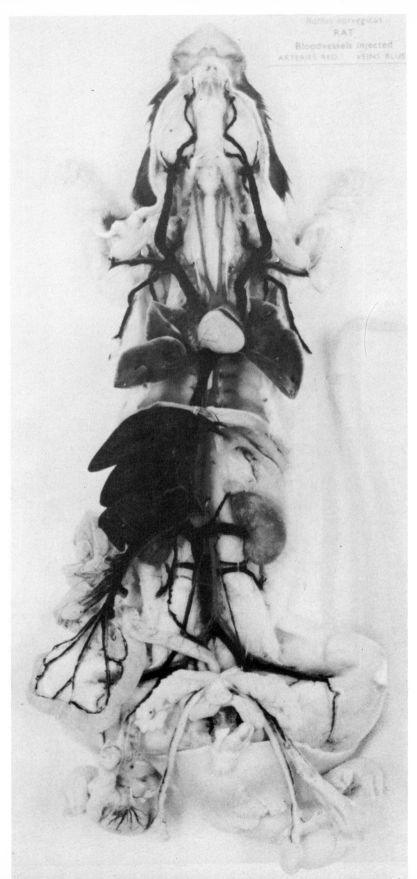

71. *Rattus*, dissection of blood vascular system. (Mag. ×1·1)

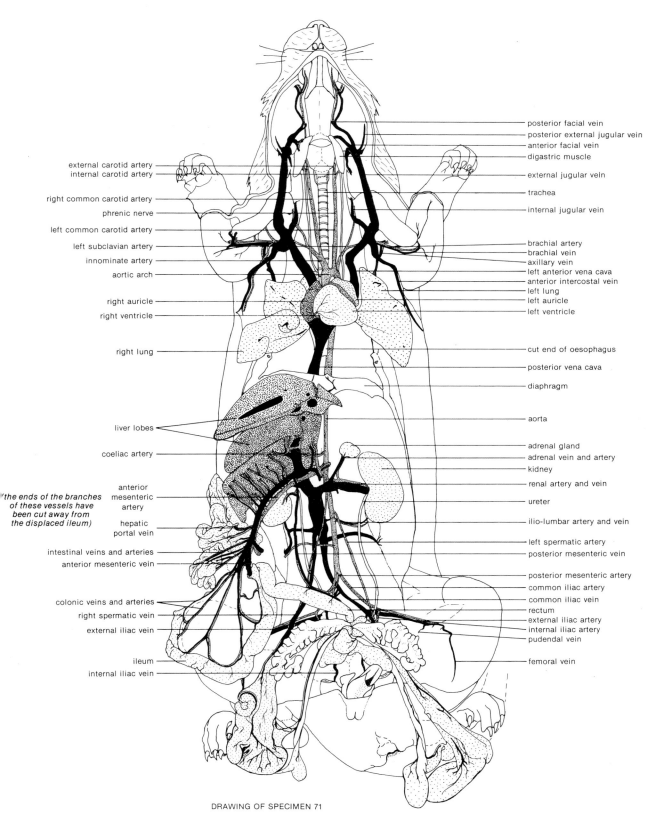

posterior facial vein
posterior external jugular vein
anterior facial vein
digastric muscle
external jugular vein
trachea
internal jugular vein

external carotid artery
internal carotid artery

right common carotid artery
phrenic nerve

left common carotid artery
left subclavian artery
innominate artery
aortic arch

brachial artery
brachial vein
axillary vein
left anterior vena cava
anterior intercostal vein
left lung
left auricle
left ventricle

right auricle
right ventricle

right lung

cut end of oesophagus

posterior vena cava

diaphragm

aorta

liver lobes

coeliac artery

adrenal gland
adrenal vein and artery
kidney
renal artery and vein
ureter
ilio-lumbar artery and vein

anterior
mesenteric
artery

(the ends of the branches
of these vessels have
been cut away from
the displaced ileum)

hepatic
portal vein

left spermatic artery
posterior mesenteric vein

intestinal veins and arteries
anterior mesenteric vein

posterior mesenteric artery
common iliac artery
common iliac vein
rectum
external iliac artery
internal iliac artery
pudendal vein

colonic veins and arteries
right spermatic vein
external iliac vein

ileum
internal iliac vein

femoral vein

DRAWING OF SPECIMEN 71

75

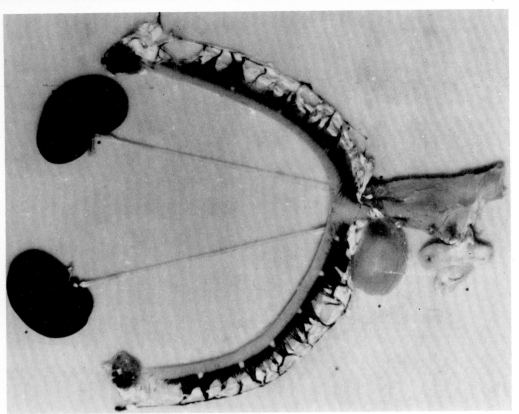

73. *Rattus*, dissection of separated female urogenital system. (Mag. ×4)

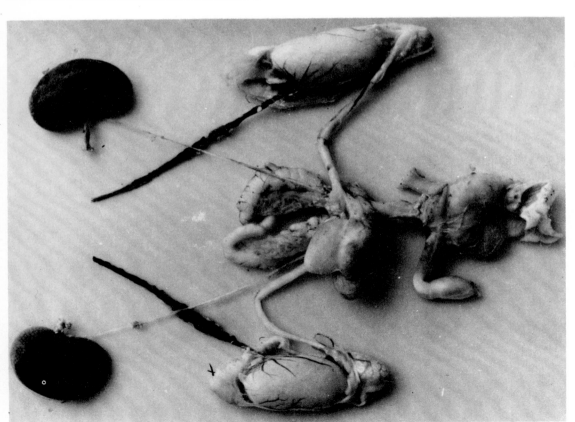

72. *Rattus*, dissection of separated male urogenital system. (Mag. ×4)

left kidney

ovary
periovarial sac
oviduct
fat body

left uterus

left ureter

embryo
uterine artery
and vein

mesentery

bladder

urethra

vagina

preputial gland

vulva
prepuce
urinary opening
rectum

DRAWING OF SPECIMEN 73

right kidney

stumps of renal artery and vein

right ureter

pampiniform plexus

fat body

caput epididymis

seminal vesicle

right vas deferens

coagulation gland

right testis

corpus epididymis

bladder

ampullary gland

lateral prostate gland

ventral prostate gland

cauda epididymis

urethra

corpus spongiosum

Cowper's gland

penis

position of
preputial gland

bulbo-cavernosum

rectum

DRAWING OF SPECIMEN 72

77

74. *Oryctolagus*, steak, head sagittal LS. (Mag. ×2)

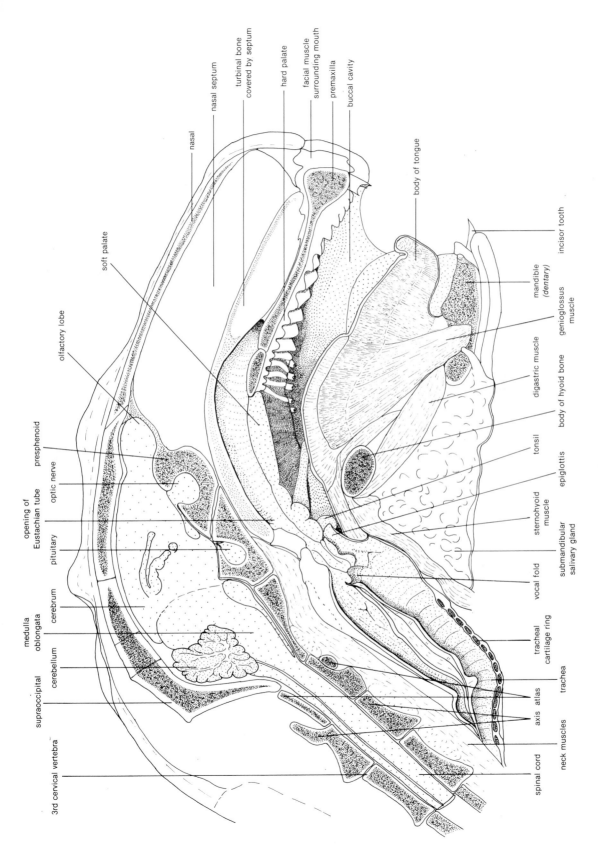

nasal

nasal septum

turbinal bone
covered by septum

hard palate

facial muscle
surrounding mouth

premaxilla

buccal cavity

body of tongue

incisor tooth

mandible
(dentary)

genioglossus
muscle

digastric muscle

body of hyoid bone

tonsil

epiglottis

sternohyoid
muscle

submandibular
salivary gland

vocal fold

tracheal
cartilage ring

trachea

axis atlas

spinal cord

neck muscles

soft palate

olfactory lobe

opening of
Eustachian tube

presphenoid

optic nerve

pituitary

cerebrum

medulla
oblongata

cerebellum

supraoccipital

3rd cervical vertebra

DRAWING OF SPECIMEN 74

79

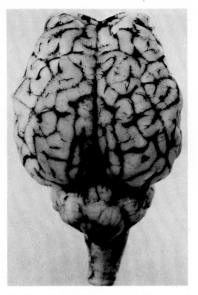

75. *Ovis*, brain, dorsal view. (Mag. ×1)

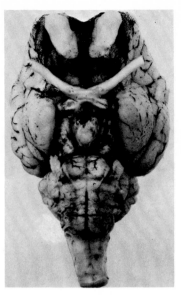

76. *Ovis*, brain, ventral view. (Mag. ×1)

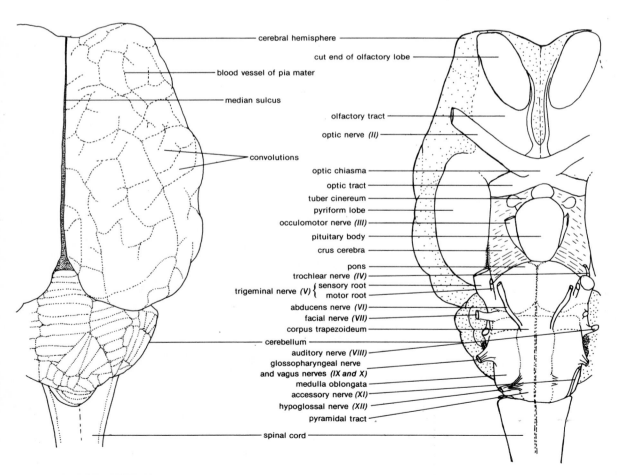

cerebral hemisphere

cut end of olfactory lobe

blood vessel of pia mater

olfactory tract

median sulcus

optic nerve *(II)*

convolutions

optic chiasma

optic tract

tuber cinereum

pyriform lobe

occulomotor nerve *(III)*

pituitary body

crus cerebra

pons

trochlear nerve *(IV)*

trigeminal nerve *(V)* { sensory root
motor root

abducens nerve *(VI)*

facial nerve *(VII)*

corpus trapezoideum

cerebellum

auditory nerve *(VIII)*

glossopharyngeal nerve
and vagus nerves *(IX and X)*

medulla oblongata

accessory nerve *(XI)*

hypoglossal nerve *(XII)*

pyramidal tract

spinal cord

DRAWING OF SPECIMEN 75

DRAWING OF SPECIMEN 76

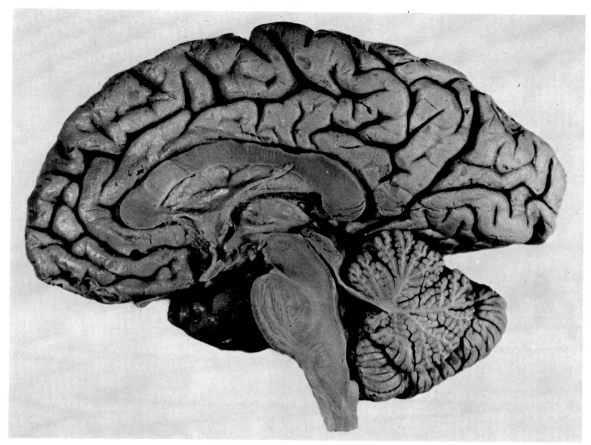

77. *Homo*, brain, median LS. (Mag. ×1)

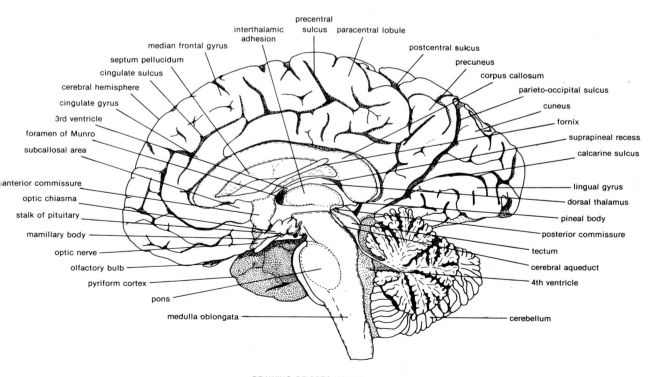

DRAWING OF SPECIMEN 77

81

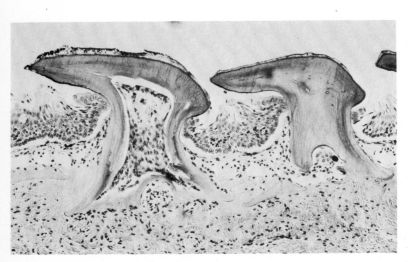

78. *Scyliorhinus*, skin VS. (Mag. ×65)

79. *Solea*, surface view of skin by reflected light. (Mag.

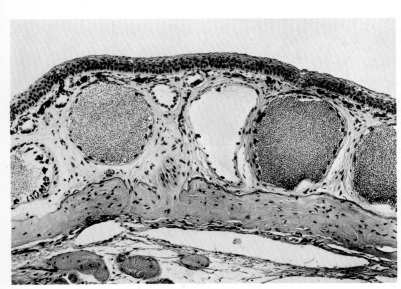

80. *Rana*, skin VS. (Mag. ×35)

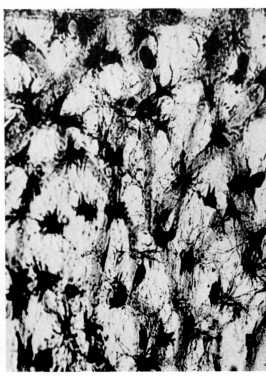

81. *Rana*, skin stretch with melanophores. (Mag. ×45)

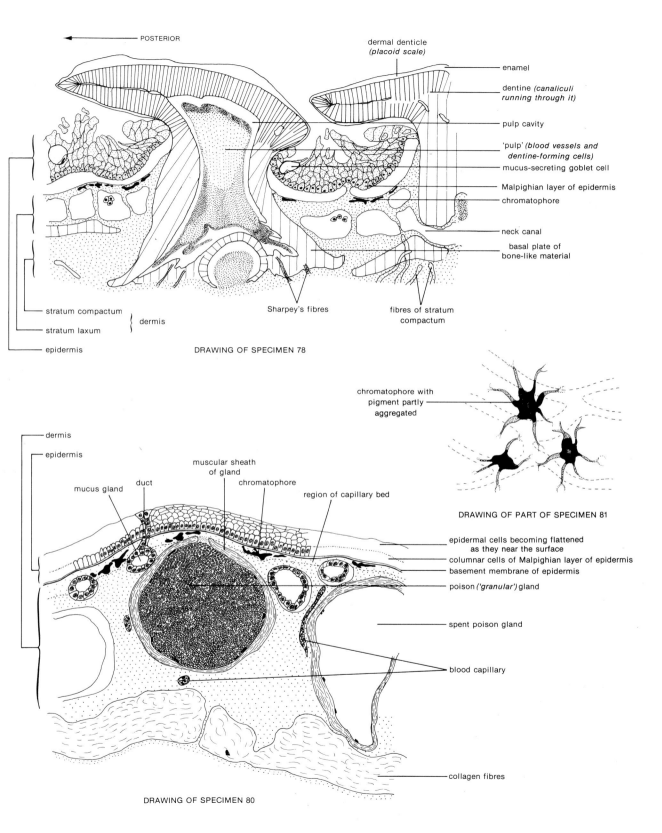

POSTERIOR

dermal denticle
(placoid scale)

enamel

dentine (canaliculi
running through it)

pulp cavity

'pulp' (blood vessels and
dentine-forming cells)

mucus-secreting goblet cell

Malpighian layer of epidermis

chromatophore

neck canal

basal plate of
bone-like material

stratum compactum

stratum laxum

dermis

epidermis

Sharpey's fibres

fibres of stratum
compactum

DRAWING OF SPECIMEN 78

chromatophore with
pigment partly
aggregated

DRAWING OF PART OF SPECIMEN 81

dermis

epidermis

muscular sheath
of gland

chromatophore

mucus gland

duct

region of capillary bed

epidermal cells becoming flattened
as they near the surface

columnar cells of Malpighian layer of epidermis

basement membrane of epidermis

poison ('granular') gland

spent poison gland

blood capillary

collagen fibres

DRAWING OF SPECIMEN 80

83

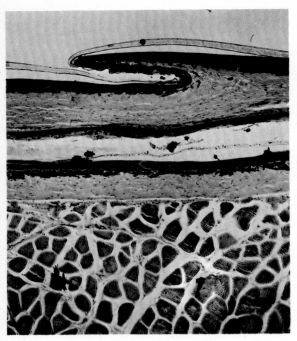

82. *Lacerta*, skin VS. (Mag. ×45)

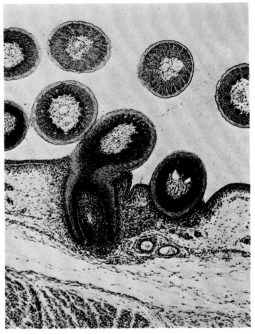

83. Bird skin, VS early feather formation. (Mag. ×65)

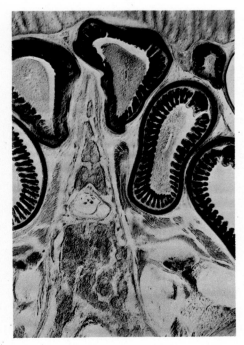

84. Bird skin, VS later feather formation. (Mag. ×60)

85. *Homo*, injected skin VS. (Mag. ×60)

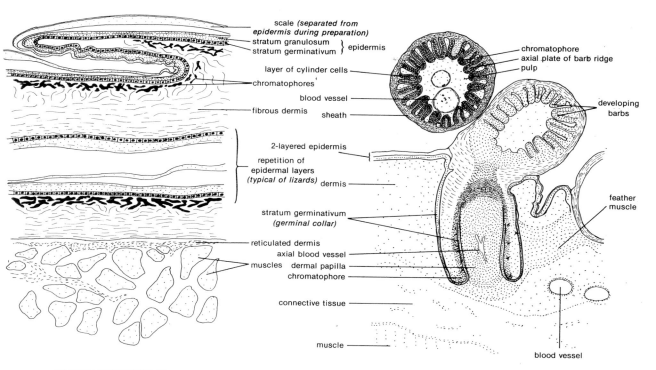

scale *(separated from epidermis during preparation)*
stratum granulosum } epidermis
stratum germinativum }
layer of cylinder cells
chromatophores'
fibrous dermis
sheath

2-layered epidermis
repetition of epidermal layers *(typical of lizards)* dermis

stratum germinativum *(germinal collar)*

reticulated dermis
axial blood vessel
muscles   dermal papilla
chromatophore

connective tissue

muscle

DRAWING OF SPECIMEN 82

chromatophore
axial plate of barb ridge
pulp

developing barbs

feather muscle

blood vessel
sheath

blood vessel

DRAWING OF SPECIMEN 83

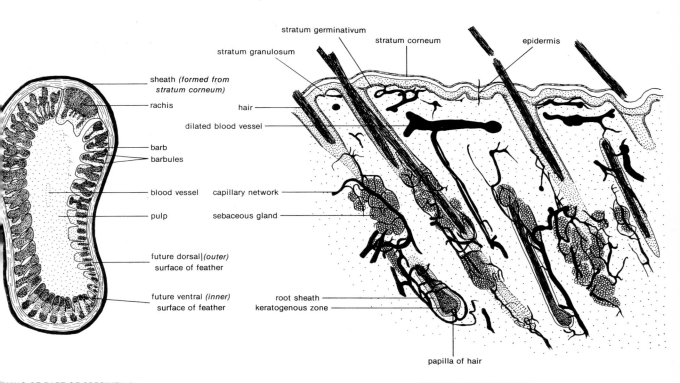

sheath *(formed from stratum corneum)*
rachis
barb
barbules

blood vessel
pulp

future dorsal|*(outer)* surface of feather

future ventral *(inner)* surface of feather

WING OF PART OF SPECIMEN 84

stratum granulosum
stratum germinativum
stratum corneum
epidermis

hair
dilated blood vessel

capillary network
sebaceous gland

root sheath
keratogenous zone

papilla of hair

DRAWING OF SPECIMEN 85

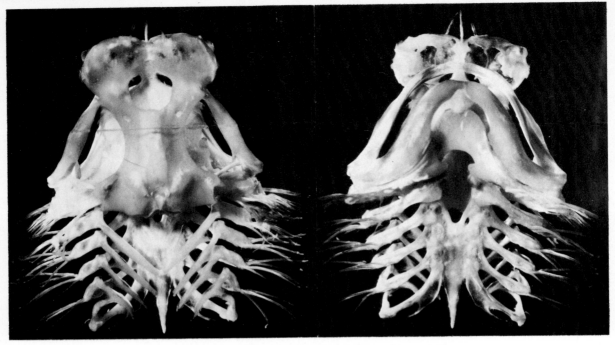

86. *Scyliorhinus*, skull and branchial skeleton. (Mag. ×0·5)

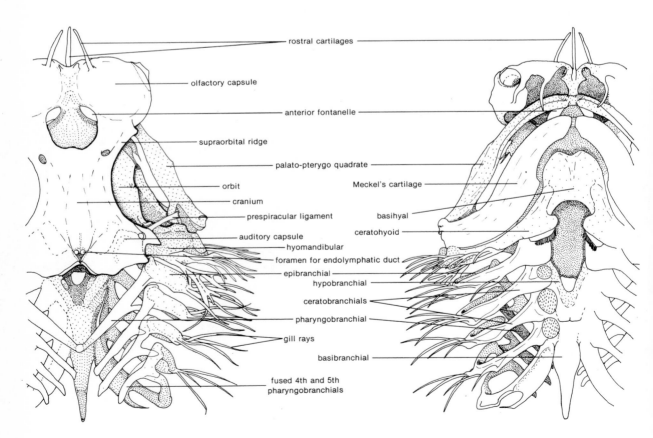

rostral cartilages

olfactory capsule

anterior fontanelle

supraorbital ridge

palato-pterygo quadrate

Meckel's cartilage

orbit

cranium

basihyal

prespiracular ligament

ceratohyoid

auditory capsule

hyomandibular

foramen for endolymphatic duct

epibranchial

hypobranchial

ceratobranchials

pharyngobranchial

gill rays

basibranchial

fused 4th and 5th
pharyngobranchials

DRAWING OF SPECIMEN 86

87. *Gadus*, skull and vertebral column. (Mag. ×0·8)

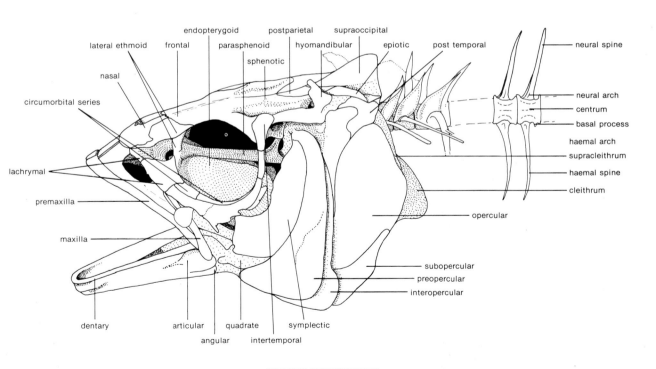

DRAWING OF SPECIMEN 87

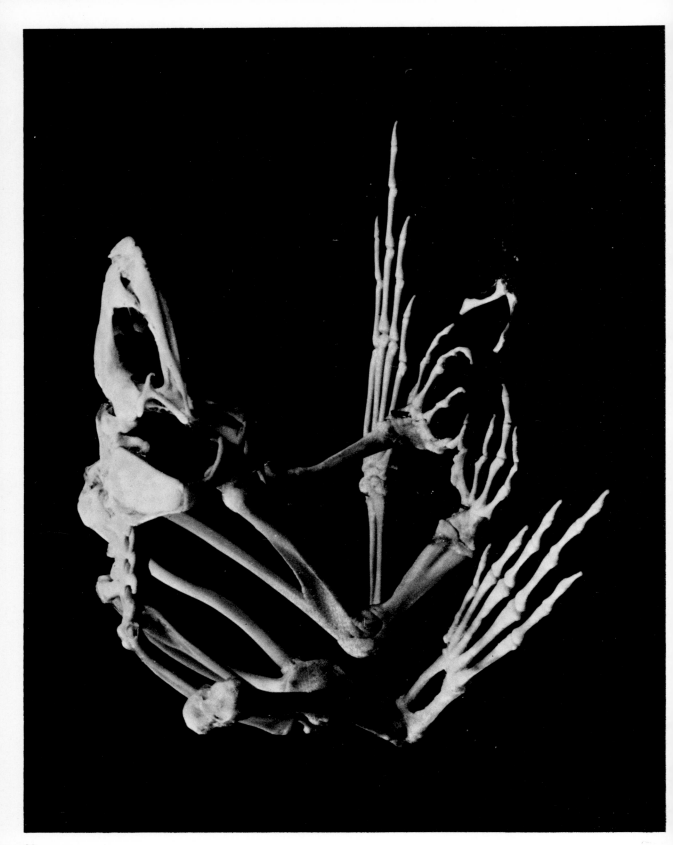

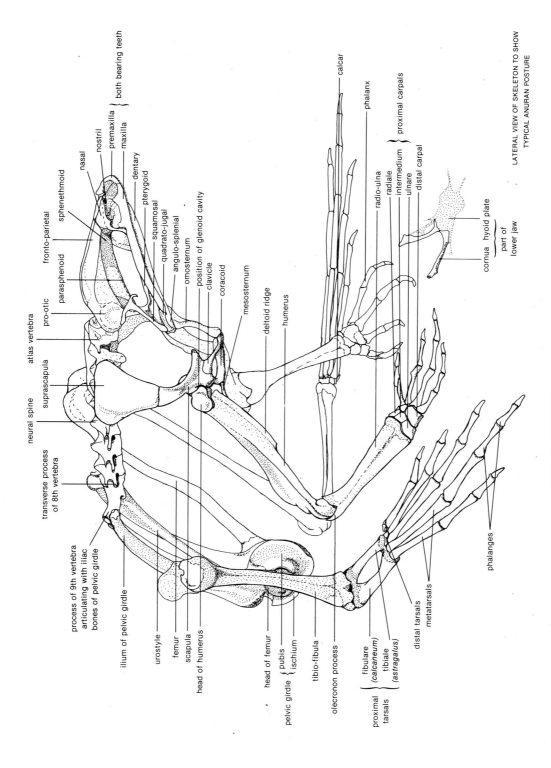

both bearing teeth

nostril
premaxilla
maxilla
nasal
dentary
sphenethmoid
pterygoid
fronto-parietal
squamosal
quadrato-jugal
parasphenoid
angulo-splenial
pro-otic
omosternum
atlas vertebra
position of glenoid cavity
neural spine
clavicle
suprascapula
coracoid
mesosternum
transverse process
of 8th vertebra
deltoid ridge
humerus

process of 9th vertebra
articulating with iliac
bones of pelvic girdle
ilium of pelvic girdle
urostyle
femur
scapula
head of humerus
head of femur
pelvic girdle { pubis
                ischium
tibio-fibula
olecronon process
proximal { fibulare
tarsals   (calcaneum)
           tibiale
           (astragalus)
distal tarsals
metatarsals
phalanges

calcar
phalanx
radio-ulna
radiale
intermedium
ulnare
distal carpal
proximal carpals

cornua hyoid plate
part of
lower jaw

LATERAL VIEW OF SKELETON TO SHOW
TYPICAL ANURAN POSTURE

DRAWING OF SPECIMEN 88

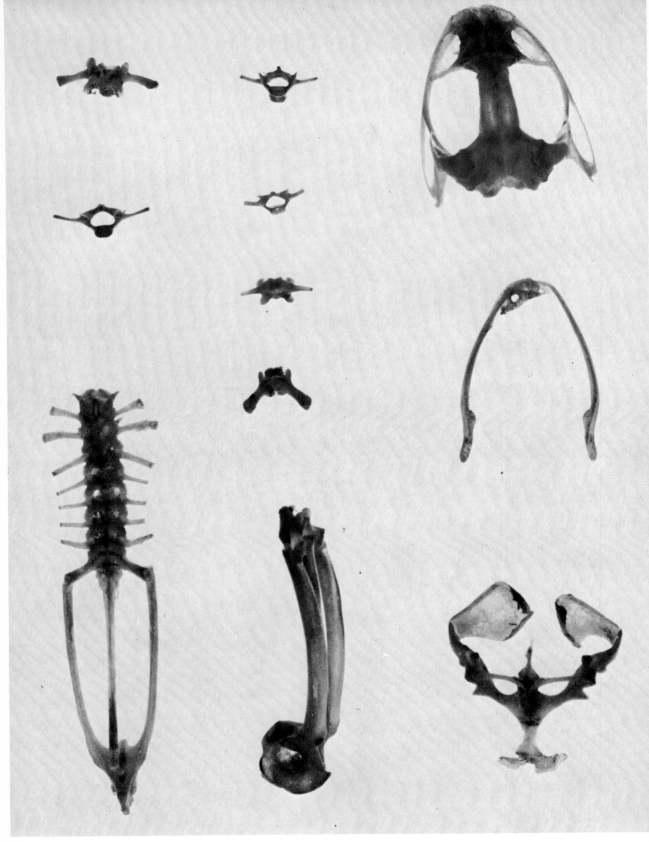

89. *Rana*, separate bones. (Mag. ×2)

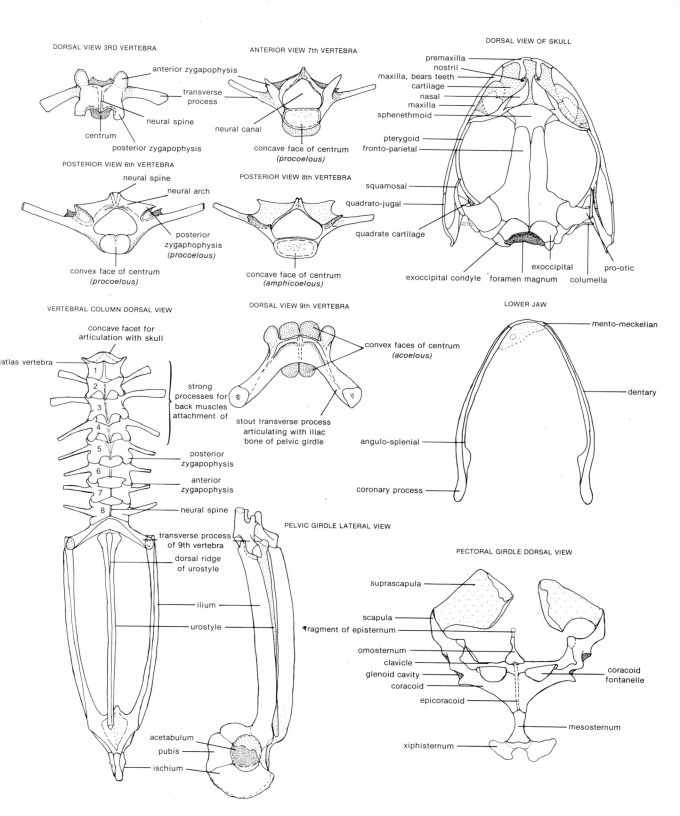

DORSAL VIEW 3RD VERTEBRA

anterior zygapophysis

transverse process

neural spine

centrum

posterior zygapophysis

ANTERIOR VIEW 7th VERTEBRA

neural canal

concave face of centrum
(*procoelous*)

DORSAL VIEW OF SKULL

premaxilla
nostril
maxilla, bears teeth
cartilage
nasal
maxilla
sphenethmoid
pterygoid
fronto-parietal

squamosal

quadrato-jugal

quadrate cartilage

exoccipital
pro-otic
exoccipital condyle   foramen magnum   columella

POSTERIOR VIEW 6th VERTEBRA

neural spine
neural arch

posterior
zygaphophysis
(*procoelous*)

convex face of centrum
(*procoelous*)

POSTERIOR VIEW 8th VERTEBRA

concave face of centrum
(*amphicoelous*)

VERTEBRAL COLUMN DORSAL VIEW

concave facet for
articulation with skull

atlas vertebra

1
2
3
4
5
6
7
8

strong
processes for
back muscles
attachment of

posterior
zygapophysis

anterior
zygapophysis

neural spine

transverse process
of 9th vertebra

dorsal ridge
of urostyle

ilium

urostyle

acetabulum
pubis
ischium

DORSAL VIEW 9th VERTEBRA

convex faces of centrum
(*acoelous*)

stout transverse process
articulating with iliac
bone of pelvic girdle

PELVIC GIRDLE LATERAL VIEW

LOWER JAW

mento-meckelian

dentary

angulo-splenial

coronary process

PECTORAL GIRDLE DORSAL VIEW

suprascapula

scapula
fragment of episternum
omosternum
clavicle
glenoid cavity
coracoid
epicoracoid

coracoid
fontanelle

mesosternum

xiphisternum

DRAWINGS FROM SPECIMEN 89

91

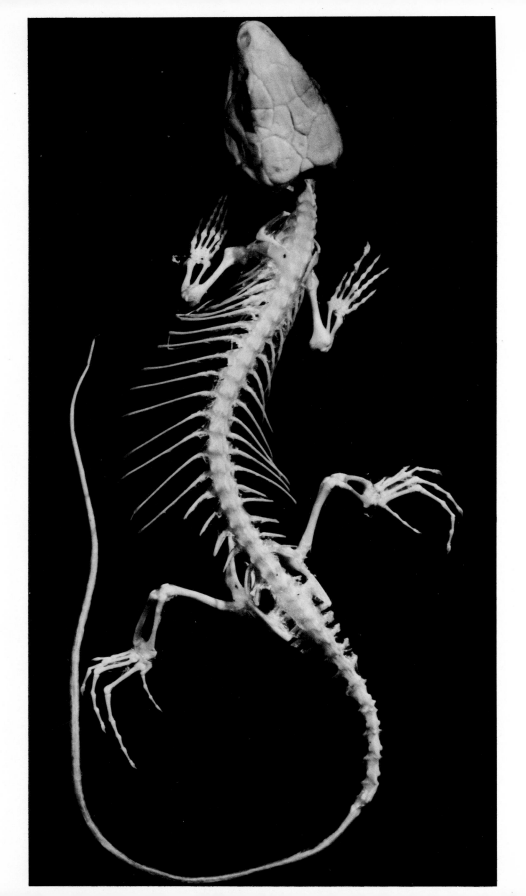

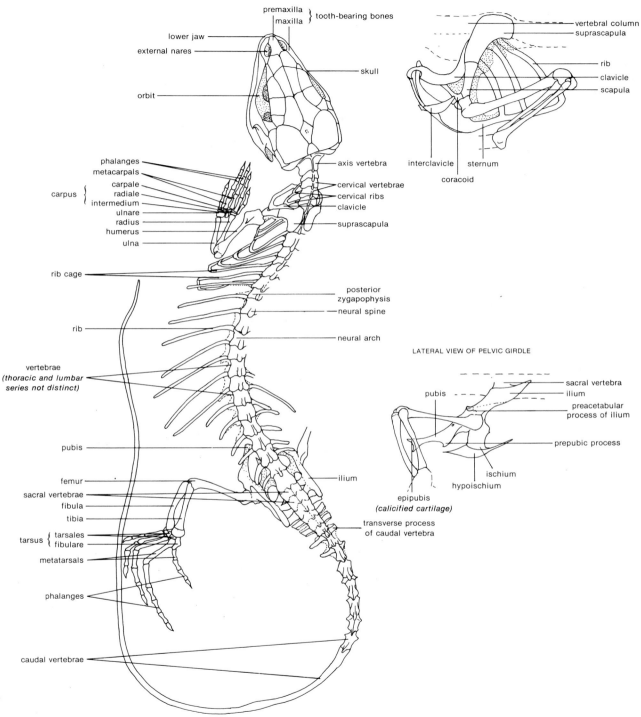

premaxilla
maxilla } tooth-bearing bones

lower jaw

external nares

orbit

skull

vertebral column
suprascapula

rib
clavicle
scapula

interclavicle   sternum

coracoid

phalanges
metacarpals
carpale
carpus {
radiale
intermedium
ulnare
radius
humerus
ulna

axis vertebra

cervical vertebrae
cervical ribs
clavicle
suprascapula

rib cage

posterior
zygapophysis
neural spine

rib

neural arch

vertebrae
*(thoracic and lumbar
series not distinct)*

LATERAL VIEW OF PELVIC GIRDLE

pubis

sacral vertebra
ilium
preacetabular
process of ilium

prepubic process

pubis

femur
sacral vertebrae
fibula
tibia
tarsales
tarsus {
fibulare
metatarsals

ilium

ischium
hypoischium

epipubis
(calicified cartilage)

transverse process
of caudal vertebra

phalanges

caudal vertebrae

Lacerta, entire skeleton. (Mag. ×1·8)

91. *Lacerta*, skull. (Mag. ×

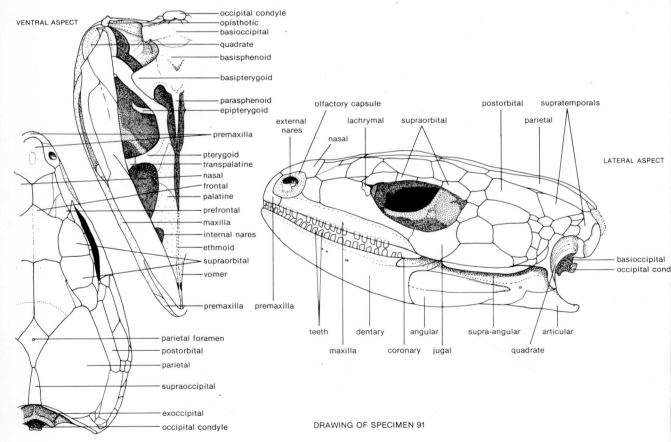

VENTRAL ASPECT

occipital condyle
opisthotic
basioccipital
quadrate
basisphenoid
basipterygoid

parasphenoid
epipterygoid

premaxilla

pterygoid
transpalatine
nasal
frontal
palatine
prefrontal
maxilla
internal nares
ethmoid
supraorbital
vomer

premaxilla

parietal foramen
postorbital
parietal

supraoccipital

exoccipital
occipital condyle

DORSAL ASPECT

olfactory capsule
external nares
nasal

lachrymal
supraorbital

postorbital
parietal

supratemporals

LATERAL ASPECT

basioccipital
occipital cond

premaxilla

teeth
maxilla

dentary

angular
coronary

supra-angular

jugal

articular

quadrate

DRAWING OF SPECIMEN 91

94

92. *Python*, skull. (Mag. ×2)

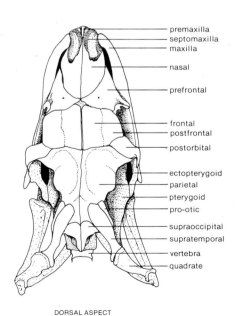

premaxilla
septomaxilla
maxilla

nasal

prefrontal

frontal
postfrontal
postorbital

ectopterygoid
parietal
pterygoid
pro-otic

supraoccipital
supratemporal
vertebra
quadrate

DORSAL ASPECT

maxilla
(bears teeth)

premaxilla   nasal

prefrontal

frontal

postorbital

parietal

ectopterygoid

supratemporal
pro-otic
basioccipital
quadrate
coronary
pterygoid

supra-angular

palatine
(bears teeth)

dentary
(bears teeth)

angular

LATERAL ASPECT

DRAWING OF SPECIMEN 92

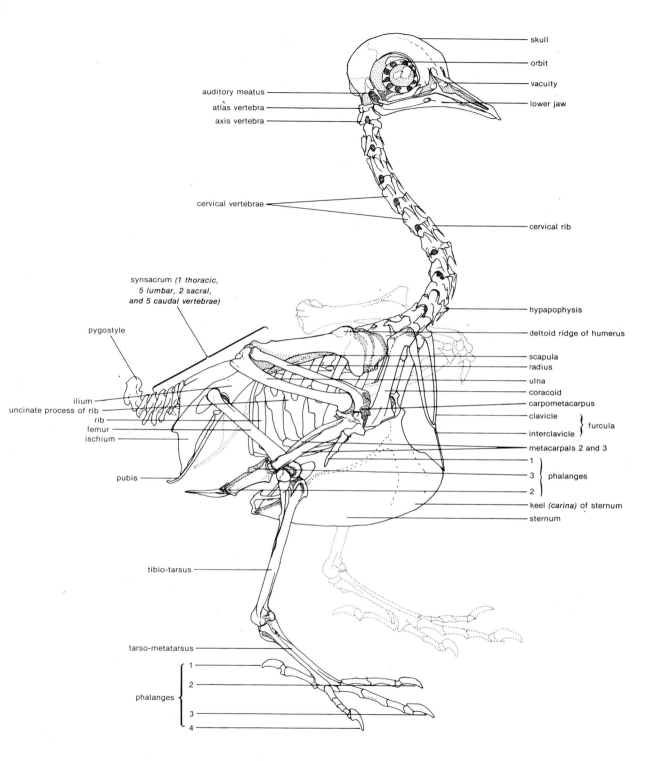

skull

orbit

vacuity

lower jaw

auditory meatus

atlas vertebra

axis vertebra

cervical vertebrae

cervical rib

synsacrum *(1 thoracic,
5 lumbar, 2 sacral,
and 5 caudal vertebrae)*

hypapophysis

deltoid ridge of humerus

pygostyle

scapula
radius
ulna
coracoid
carpometacarpus
clavicle
interclavicle } furcula

ilium
uncinate process of rib
rib
femur
ischium

metacarpals 2 and 3

1
3 } phalanges
2

pubis

keel *(carina)* of sternum
sternum

tibio-tarsus

tarso-metatarsus

1
2
phalanges {
3
4

DRAWING OF SPECIMENT 93

Columba, entire skeleton. (Mag. ×0·6)

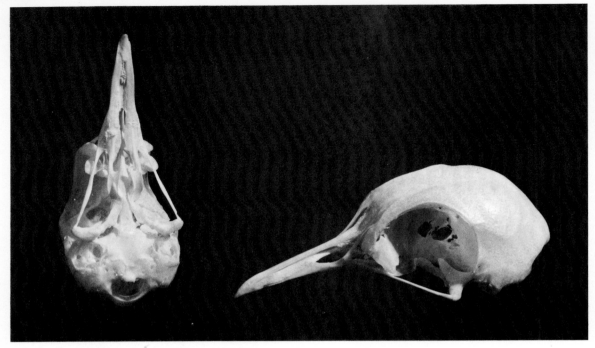

94. *Columba*, skull. (Mag. ×2)

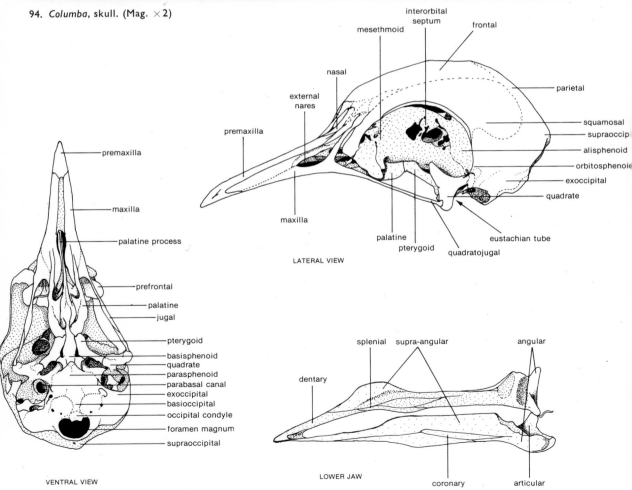

premaxilla

maxilla

palatine process

prefrontal

palatine

jugal

pterygoid

basisphenoid

quadrate

parasphenoid

parabasal canal

exoccipital

basioccipital

occipital condyle

foramen magnum

supraoccipital

VENTRAL VIEW

interorbital septum

mesethmoid

frontal

nasal

external nares

parietal

premaxilla

squamosal

supraoccip

alisphenoid

orbitosphenoi

exoccipital

quadrate

maxilla

palatine

pterygoid

quadratojugal

eustachian tube

LATERAL VIEW

splenial

supra-angular

angular

dentary

coronary

articular

LOWER JAW

98   DRAWING OF SPECIMEN 94

Columba, wing. (Mag. ×0·8)

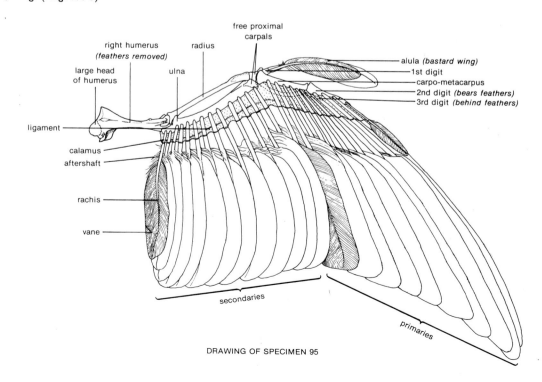

large head
of humerus

right humerus
(feathers removed)

radius

free proximal
carpals

ulna

alula *(bastard wing)*

1st digit

carpo-metacarpus

2nd digit *(bears feathers)*

3rd digit *(behind feathers)*

ligament

calamus

aftershaft

rachis

vane

secondaries

primaries

DRAWING OF SPECIMEN 95

96. *Oryctolagus*, entire skeleton. (Mag. ×0·6)

frontal
intraorbital foramen
external auditory meatus
zygomatic arch
nasal
premaxilla
diastema
maxilla
incisor tooth
premolar
molars
mandible

axis vertebra
tympanic bulla
parietal
1st cervical vertebra (atlas)

angular process
sternal ribs
coraco-scapula
clavicle
manubrium (1st sternebra)
xiphisternum (7th sternebra)
head of humerus
greater tubercle
deltoid ridge
humerus
condyle
sigmoid notch
ulna
radius

pectoral girdle

3rd to 7th cervical vertebrae

anterior thoracic ribs

capitulum

tuberculum

posterior thoracic vertebrae

false ribs

floating ribs
sternebrae
patella
patellar groove
condyle
femur

xiphoid cartilage
olecranon process

radiale
intermedium
centrale
trapezium
magnum
ulnare
unciform

metacarpals

phalanx
claw

cnemial crest
fibula
tibia
3rd trochanter
head of femur
greater trochanter

phalanges
claw

metatarsals
epiphyses

mesocuneiform
ectocuneiform
cuboid

tarsals

fibulare (calcaneum)

lumbar vertebrae

caudal vertebrae
ilium

sacrum

obturator foramen
pubis
ischium
ischial tuberosity

pelvic girdle

centrale
tibiale (astragalus)

DRAWING OF SPECIMEN 96

101

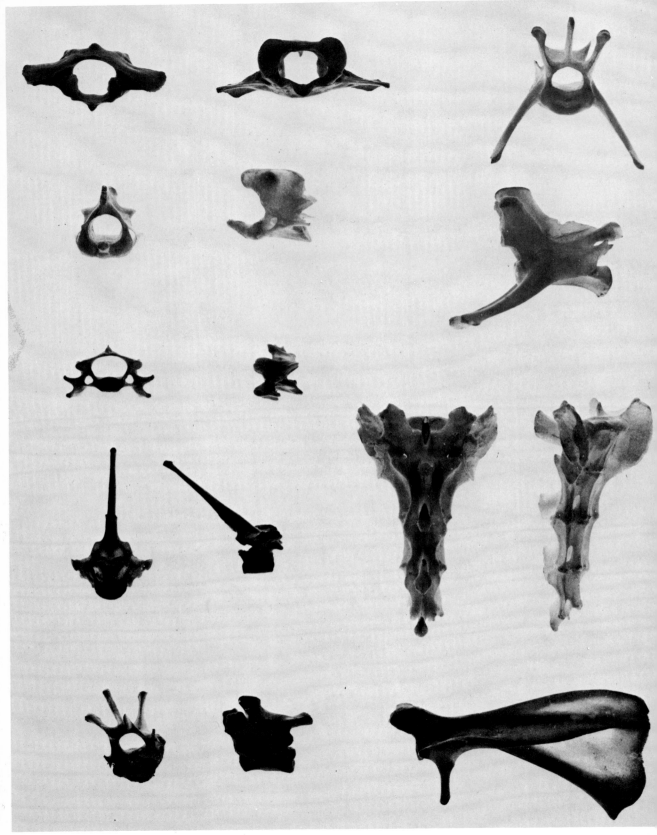

97. *Oryctolagus*, separate bones. (Mag. ×0·7)

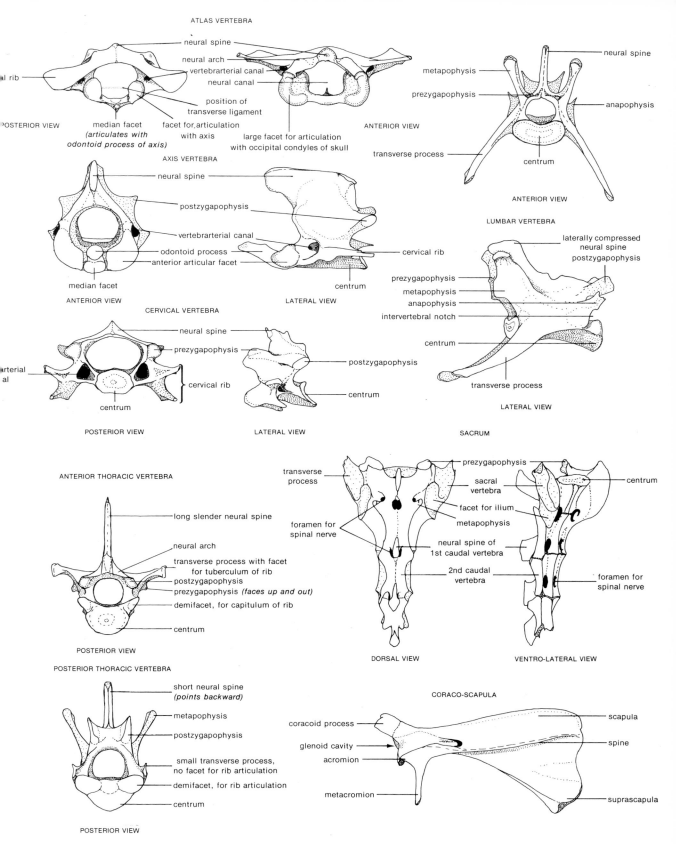

ATLAS VERTEBRA

neural spine
neural arch
vertebrarterial canal
neural canal
position of transverse ligament

al rib

POSTERIOR VIEW

median facet
(articulates with odontoid process of axis)

facet for articulation with axis

large facet for articulation with occipital condyles of skull

ANTERIOR VIEW

neural spine
metapophysis
prezygapophysis
anapophysis
transverse process
centrum

ANTERIOR VIEW

AXIS VERTEBRA

neural spine
postzygapophysis
vertebrarterial canal
odontoid process
anterior articular facet

cervical rib

centrum

median facet

ANTERIOR VIEW

LATERAL VIEW

LUMBAR VERTEBRA

laterally compressed neural spine
postzygapophysis
prezygapophysis
metapophysis
anapophysis
intervertebral notch
centrum

transverse process

LATERAL VIEW

CERVICAL VERTEBRA

neural spine
prezygapophysis
postzygapophysis
centrum

arterial al

cervical rib

centrum

POSTERIOR VIEW

LATERAL VIEW

SACRUM

ANTERIOR THORACIC VERTEBRA

long slender neural spine
neural arch
transverse process with facet for tuberculum of rib
postzygapophysis
prezygapophysis (faces up and out)
demifacet, for capitulum of rib
centrum

POSTERIOR VIEW

transverse process
foramen for spinal nerve

prezygapophysis
sacral vertebra
facet for ilium
metapophysis
neural spine of 1st caudal vertebra
2nd caudal vertebra
centrum
foramen for spinal nerve

DORSAL VIEW

VENTRO-LATERAL VIEW

POSTERIOR THORACIC VERTEBRA

short neural spine (points backward)
metapophysis
postzygapophysis
small transverse process, no facet for rib articulation
demifacet, for rib articulation
centrum

POSTERIOR VIEW

CORACO-SCAPULA

coracoid process
glenoid cavity
acromion
metacromion

scapula
spine
suprascapula

DRAWING FROM SPECIMEN 97

103

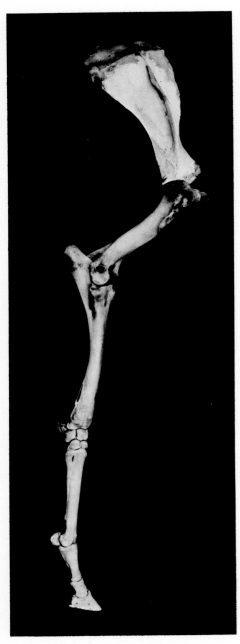

98. *Equus*, forelimb skeleton. (Mag. $\times$ 0·2)

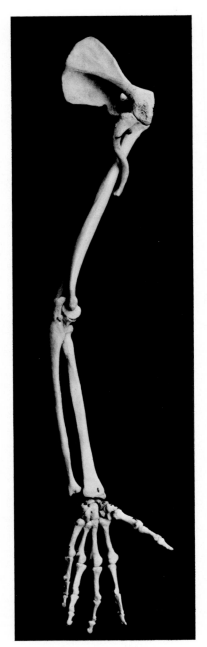

99. *Homo*, forelimb skeleton. (Mag. 0·3)

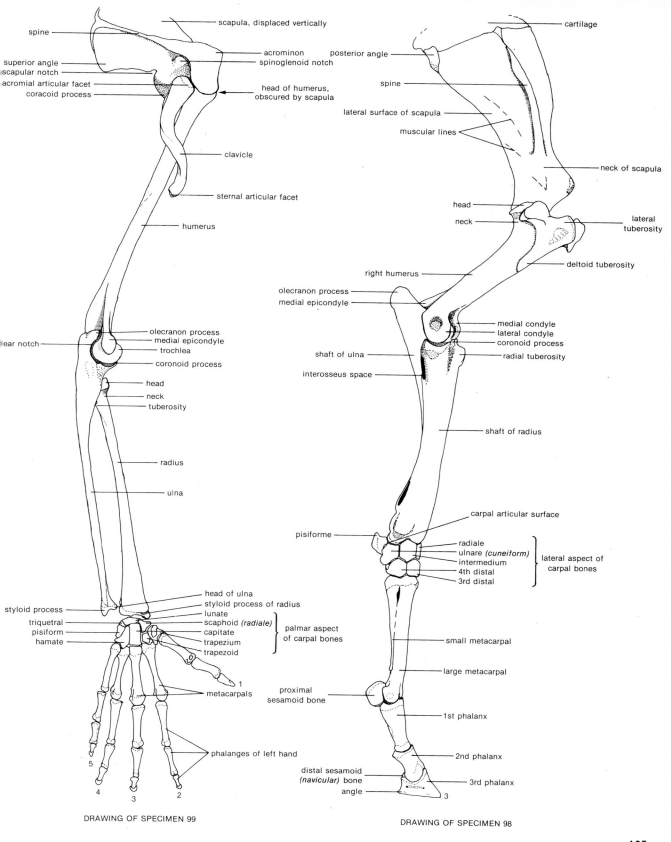

spine

superior angle

scapular notch

acromial articular facet

coracoid process

scapula, displaced vertically

acrominon

spinoglenoid notch

head of humerus,
obscured by scapula

clavicle

sternal articular facet

humerus

olecranon process

medial epicondyle

trochlea

coronoid process

ear notch

head

neck

tuberosity

radius

ulna

head of ulna

styloid process of radius

lunate

styloid process

triquetral

pisiform

hamate

scaphoid (radiale)

capitate

trapezium

trapezoid

palmar aspect
of carpal bones

1

metacarpals

proximal
sesamoid bone

5

4   3   2

phalanges of left hand

DRAWING OF SPECIMEN 99

cartilage

posterior angle

spine

lateral surface of scapula

muscular lines

neck of scapula

head

neck

lateral
tuberosity

deltoid tuberosity

right humerus

olecranon process

medial epicondyle

medial condyle

lateral condyle

coronoid process

radial tuberosity

shaft of ulna

interosseus space

shaft of radius

carpal articular surface

pisiforme

radiale

ulnare (cuneiform)

intermedium

4th distal

3rd distal

lateral aspect of
carpal bones

small metacarpal

large metacarpal

1st phalanx

2nd phalanx

distal sesamoid
(navicular) bone

angle

3rd phalanx

3

DRAWING OF SPECIMEN 98

105

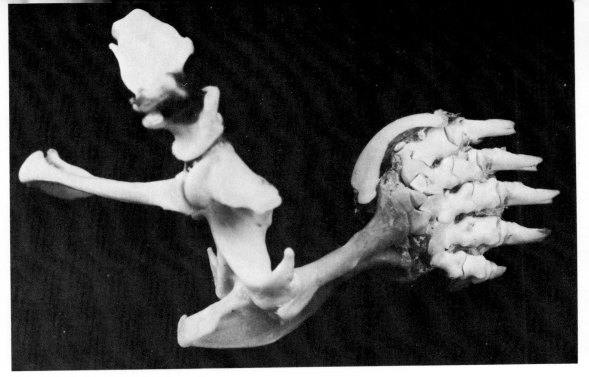

100. *Talpa*, forelimb skeleton. (Mag. ×4)

101. Bat, forelimb skeleton. (Mag. ×0·9)

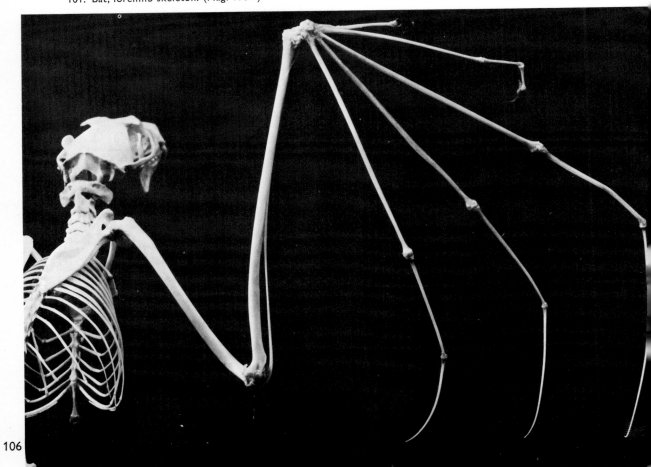

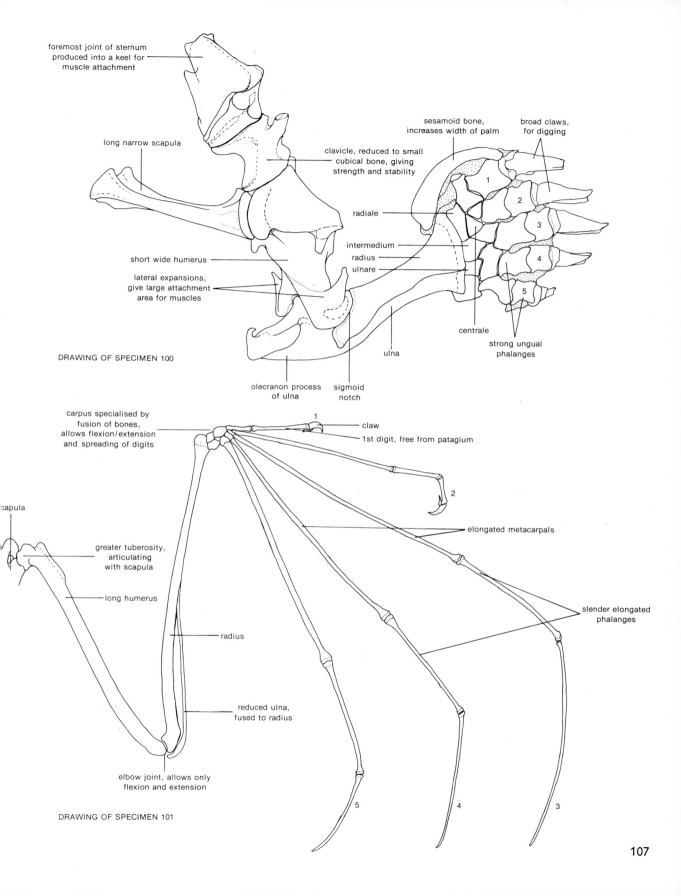

foremost joint of sternum produced into a keel for muscle attachment

long narrow scapula

clavicle, reduced to small cubical bone, giving strength and stability

sesamoid bone, increases width of palm

broad claws, for digging

radiale

1

2

3

4

5

short wide humerus

intermedium

radius

ulnare

lateral expansions, give large attachment area for muscles

centrale

strong ungual phalanges

ulna

DRAWING OF SPECIMEN 100

olecranon process of ulna

sigmoid notch

carpus specialised by fusion of bones, allows flexion/extension and spreading of digits

1

claw

1st digit, free from patagium

2

elongated metacarpals

scapula

greater tuberosity, articulating with scapula

long humerus

slender elongated phalanges

radius

reduced ulna, fused to radius

elbow joint, allows only flexion and extension

5

4

3

DRAWING OF SPECIMEN 101

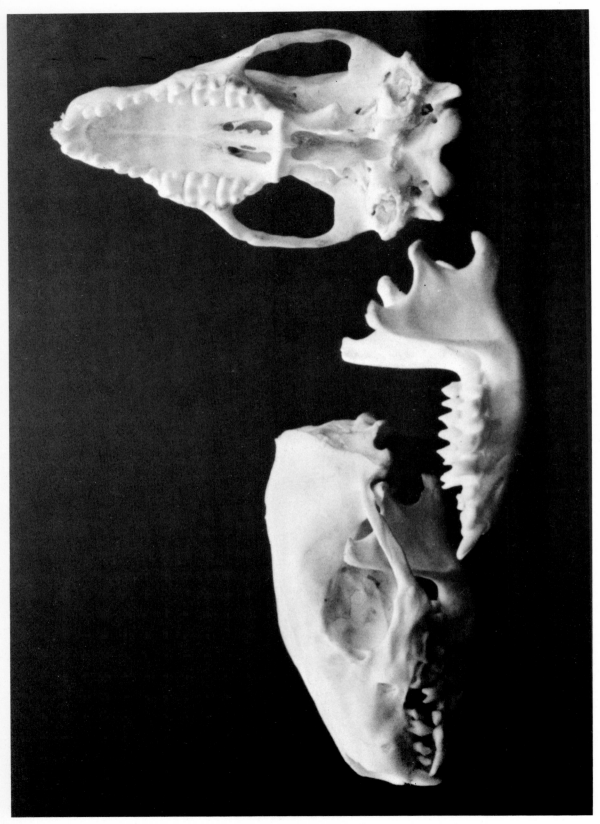

102. *Erinaceus*, skull. (Mag. ×1·8)

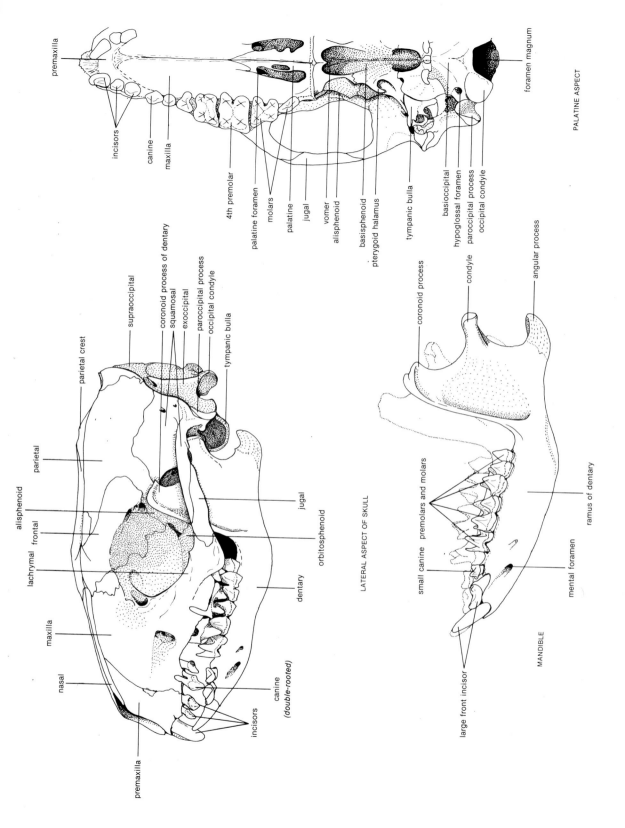

premaxilla

incisors

canine

maxilla

4th premolar

palatine foramen

molars

palatine

jugal

vomer

alisphenoid

basisphenoid

pterygoid halamus

tympanic bulla

basioccipital

hypoglossal foramen

paroccipital process

occipital condyle

foramen magnum

PALATINE ASPECT

supraoccipital

coronoid process of dentary

squamosal

exoccipital

paroccipital process

occipital condyle

tympanic bulla

parietal crest

parietal

alisphenoid

frontal

lachrymal

maxilla

nasal

premaxilla

incisors

canine
(double-rooted)

jugal

orbitosphenoid

dentary

LATERAL ASPECT OF SKULL

coronoid process

condyle

angular process

small canine

premolars and molars

large front incisor

ramus of dentary

mental foramen

MANDIBLE

DRAWING OF SPECIMEN 102

109

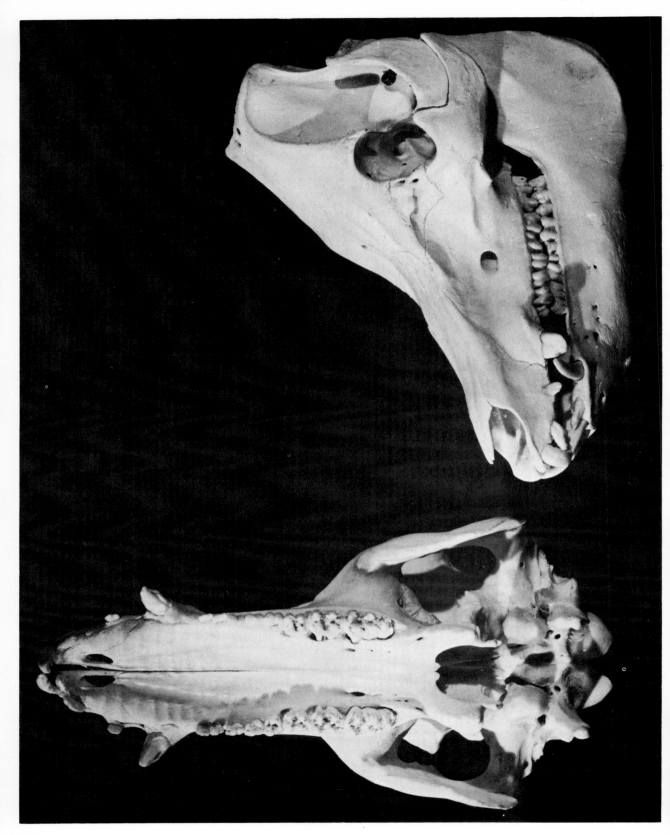

supraoccipital

external auditory meatus

lachrymal

jugal

infraorbital foramen

angle of mandible

condyle

vertical ramus

coronoid process

DETAIL OF ARTICULAR REGION OF MANDIBLE

nuchal crest

transverse crest

parietal

squamous region of temporal

supraorbital foramen and groove

frontal

zygomatic process of temporal

premaxilla

molars

premolars

mental foramina

mental prominence

body of mandible

incisors

canine

maxilla

turbinal bones

nasal

LATERAL VIEW

DRAWING OF SPECIMEN 103

incisors

palatine fissure

premaxilla

canine

palatine groove in maxilla

premolars

molars

NOTE
48 teeth
are present

maxilla

anterior palatine foramen

palatine

zygomatic arch

pterygoid process of palatine

vomer

pterygoid hamulus

pterygoid

sphenoid

temporal condyle

foramen lacerum, anterior

tympanic bulla

foramen lacerum, posterior

hypoglossal foramen

basilar part of occipital

paramastoid process

squamous part of occipital

foramen magnum

occipital condyle

VENTRAL VIEW

111

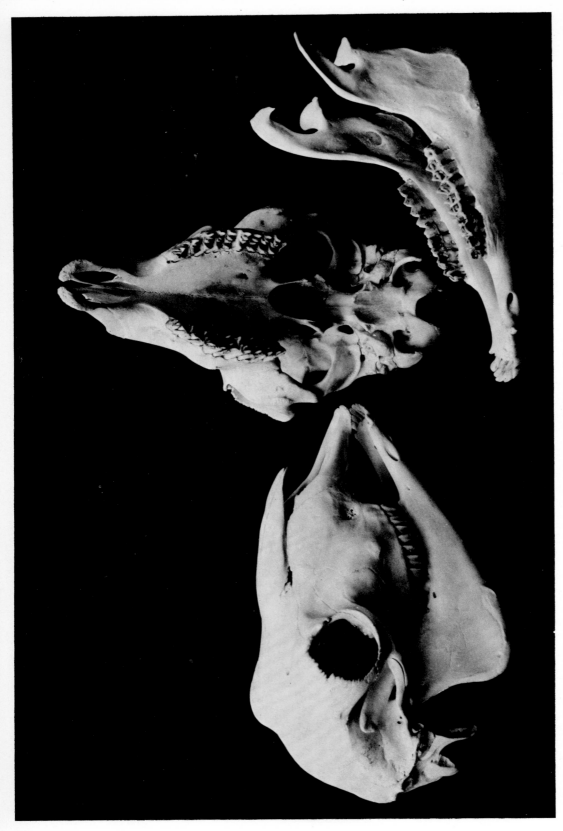

104. *Ovis*, skull. (Mag. ×0·6)

VENTRAL ASPECT OF CRANIUM

premaxilla, bearing
horny pad

internasal septum

maxilla

selenodont premolars

palatine
frontal
jugal
vomer
alisphenoid
presphenoid
squamosal
basisphenoid
foramen ovale
condyle
basioccipital
tympanic
bulla
paroccipital process
foramen magnum
occipital
condyle
exoccipital

ramus of dentary

insertion area for
masseter muscle

angular process

LOWER JAW (MANDIBLE)

nasal

turbinals

premaxilla

maxilla

jugal

palatine

squamosal

external auditory meatus

tympanic bulla

dentary

coronoid process

mandibular foramen

cheek teeth { molars
premolars

diastema

body of dentary

mental
foramen

symphisis

incisors

postorbital bar

frontal

alisphenoid

lachrymal

parietal

coronoid process
of dentary

supraoccipital

exoccipital

otic

occipital
condyle

paroccipital
process

LATERAL ASPECT OF SKULL

DRAWING OF SPECIMEN 104

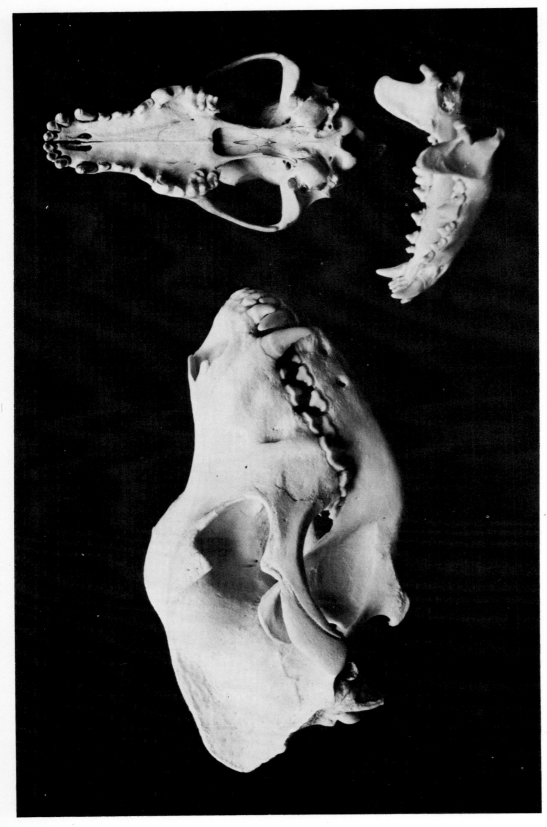

105. *Canis*, skull. (Mag. × 0·6)

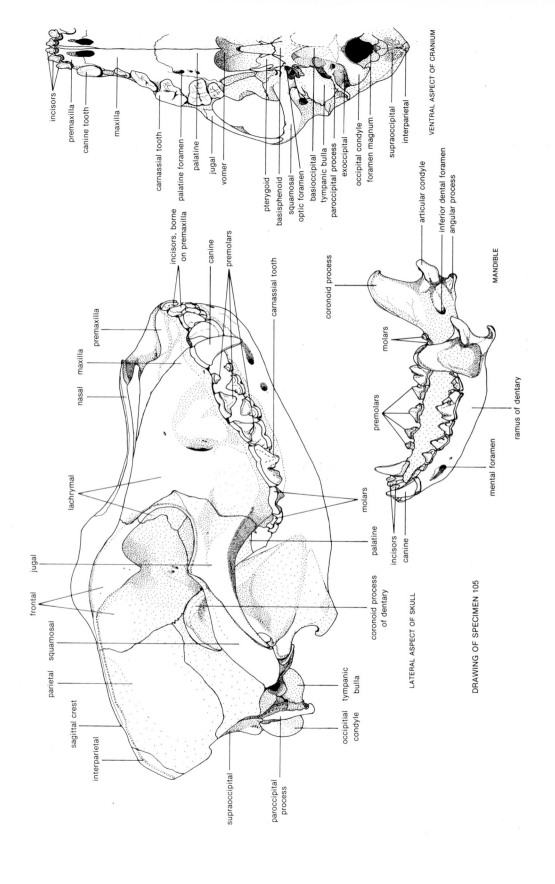

incisors
premaxilla
canine tooth
maxilla

carnassial tooth
palatine foramen
palatine
jugal
vomer

pterygoid
basisphenoid
squamosal
optic foramen
basioccipital
tympanic bulla
paroccipital process
exoccipital
occipital condyle
foramen magnum
supraoccipital
interparietal

VENTRAL ASPECT OF CRANIUM

articular condyle
inferior dental foramen
angular process

MANDIBLE

incisors, borne
on premaxilla
canine
premolars

carnassial tooth

coronoid process

molars

premaxilla
maxilla
nasal

lachrymal

frontal
jugal

parietal squamosal

sagittal crest

interparietal

supraoccipital

paroccipital
process

occipital
condyle

tympanic
bulla

coronoid process
of dentary

molars

palatine

incisors
canine

premolars

ramus of dentary

mental foramen

LATERAL ASPECT OF SKULL

DRAWING OF SPECIMEN 105

115

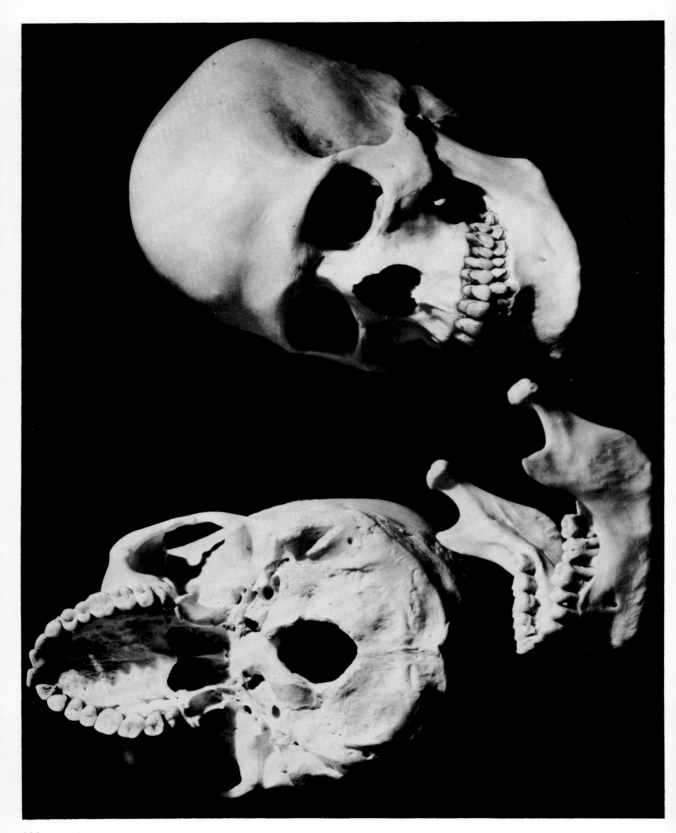

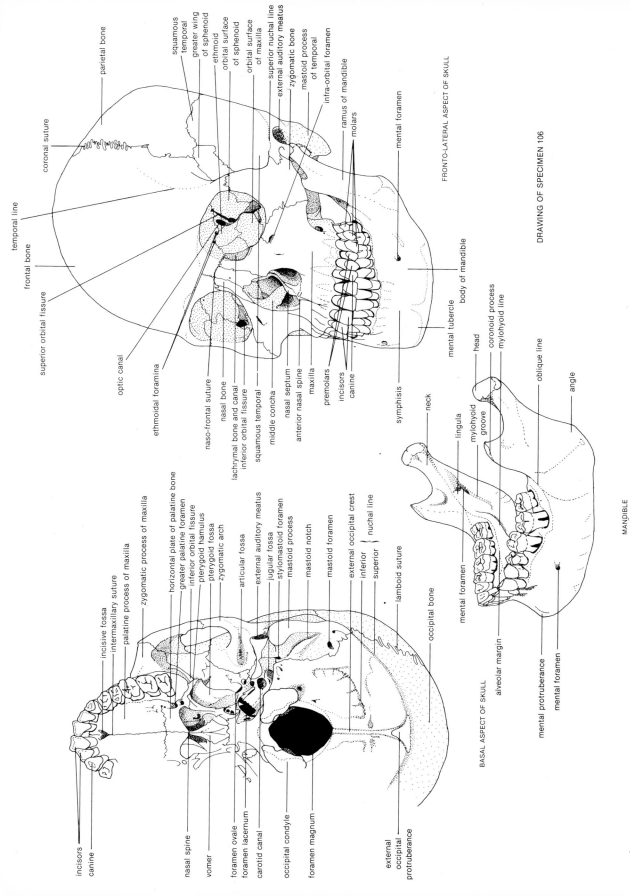

parietal bone

coronal suture

temporal line

frontal bone

superior orbital fissure

optic canal

ethmoidal foramina

squamous temporal
greater wing of sphenoid
ethmoid
orbital surface of sphenoid
orbital surface of maxilla
superior nuchal line
external auditory meatus
zygomatic bone
mastoid process of temporal
infra-orbital foramen

ramus of mandible

molars

mental foramen

naso-frontal suture
nasal bone
lachrymal bone and canal
inferior orbital fissure
squamous temporal
middle concha
nasal septum
anterior nasal spine
maxilla
premolars
incisors
canine

symphisis

mental tubercle

body of mandible

FRONTO-LATERAL ASPECT OF SKULL

DRAWING OF SPECIMEN 106

incisors
canine

nasal spine

vomer

foramen ovale
foramen lacernum
carotid canal
occipital condyle
foramen magnum

external
occipital
protruberance

incisive fossa
intermaxillary suture
palatine process of maxilla
zygomatic process of maxilla
horizontal plate of palatine bone
greater palatine foramen
inferior orbital fissure
pterygoid hamulus
pterygoid fossa
zygomatic arch

articular fossa

external auditory meatus
jugular fossa
stylomastoid foramen
mastoid process
mastoid notch

mastoid foramen

external occipital crest
inferior } nuchal line
superior }

lamboid suture

occipital bone

BASAL ASPECT OF SKULL

neck

lingula

mylohyoid groove

head

coronoid process

mylohyoid line

oblique line

angle

alveolar margin

mental foramen

mental protruberance

mental foramen

MANDIBLE

117

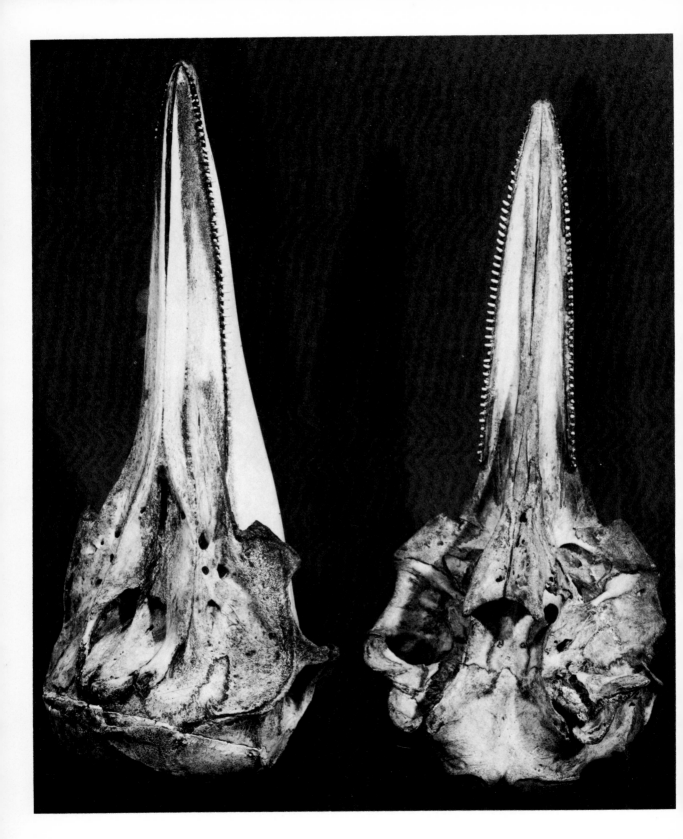

typical of the Odontoceti

DORSO-LATERAL ASPECT

premaxilla

conical peg-like
homodont teeth

maxilla

forward extension of
skull into rostrum

external nares leading to nasal
passage lacking turbinal bones

straight ramus
of mandible

extension of premaxilla
(left shorter than right)

zygomatic
process

squamosal

nasal

supraoccipital

frontal, largely covered by
extension of maxilla

interparietal

parietal

occipital condyle

paroccipital process

DETAIL OF ARTICULAR END OF MANDIBLE

coronoid process

large flat articular process

angular process

maxilla

VENTRAL ASPECT

palatine

pterygoid

internal nares

basisphenoid

basioccipital

articular surface
of squamosal
(glenoid fossa)

tympanic bulla

external auditory meatus

occipital condyle

foramen magnum

DRAWING OF SPECIMEN 107